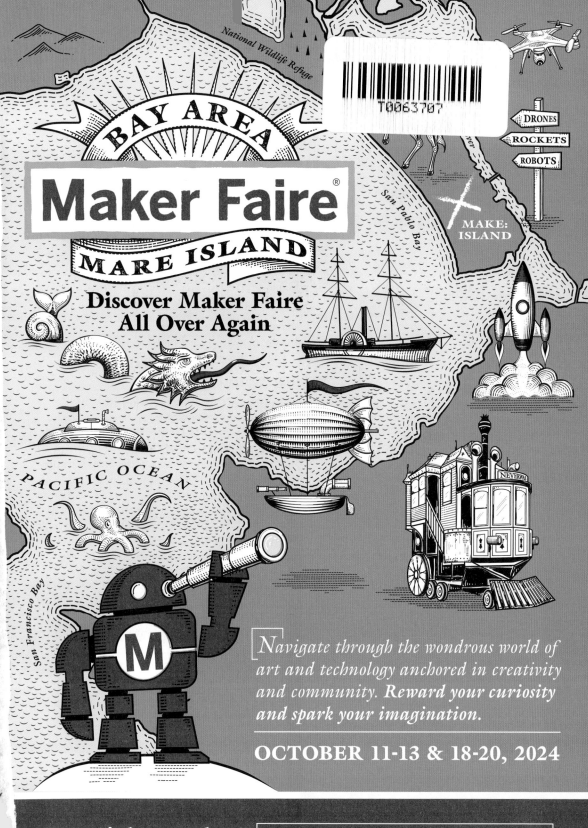

BAY AREA
Maker Faire®
MARE ISLAND

Discover Maker Faire All Over Again

National Wildlife Refuge

T0063707

DRONES
ROCKETS
ROBOTS

San Pablo Bay

MAKE: ISLAND

San Francisco Bay

PACIFIC OCEAN

NEVERWAS

[N]avigate through the wondrous world of art and technology anchored in creativity and community. Reward your curiosity and spark your imagination.

OCTOBER 11-13 & 18-20, 2024

Make: 88 CONTENTS

28

ON THE COVER:
Multicolor and multi-material 3D printing are more approachable for makers than ever before.

Photos:
• Bambu X1-Carbon printer — Erin Fezell and Jeremy Kolonay. (3D model and print of KC Green's *This Is Fine* cartoon by Erin Fezell.)
• Pixel Clock Plus — Charlyn Gonda

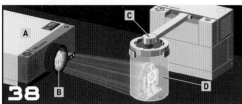

Courtney Blum, Caleb Kraft, Rob Nance, Claire Danielle Cassidy, Debra Ansell, David Battino, Nanoe

Make:

> "When I make, and I pay attention to the things I'm making, I learn important things about myself, and about being a better person."
> —Adam Savage

PRESIDENT
Dale Dougherty
dale@make.co

VP, PARTNERSHIPS
Todd Sotkiewicz
todd@make.co

EDITORIAL

EDITOR-IN-CHIEF
Keith Hammond
keith@make.co

SENIOR EDITOR
Caleb Kraft
caleb@make.co

COMMUNITY EDITOR
David J. Groom
david@make.co

PRODUCTION MANAGER
Craig Couden

CONTRIBUTING EDITORS
Tim Deagan
William Gurstelle

CONTRIBUTING WRITERS
Debra Ansell, David Battino, Courtney Blum, Rich Cameron, Claire Danielle Cassidy, Ben Eadie, Denzel Edwards, Elyse Farris, Andy Forest, Charlyn Gonda, Paul J. Henley, Joan Horvath, Bob Knetzger, Joshua Pearce, Marshall Piros, Reade Richard, Sujay Saravanan, Andrea Schick, Nick Schick, Brenda Shivanandan, Joshua R. Taylor, Taylor Waddell, Lee Wilkins, Matt Zigler

CONTRIBUTING ARTISTS
Erin Fezell, KC Green, Jeremy Kolonay, Rob Nance

MAKE.CO

ENGINEERING MANAGER
Alicia Williams

WEB APPLICATION DEVELOPER
Rio Roth-Barreiro

DESIGN

CREATIVE DIRECTOR
Juliann Brown

BOOKS

BOOKS EDITOR
Kevin Toyama
books@make.co

GLOBAL MAKER FAIRE

MANAGING DIRECTOR, GLOBAL MAKER FAIRE
Katie D. Kunde

GLOBAL LICENSING
Jennifer Blakeslee

MARKETING

DIRECTOR OF MARKETING
Gillian Mutti

PROGRAM COORDINATOR
Jamie Agius

OPERATIONS

ADMINISTRATIVE MANAGER
Cathy Shanahan

ACCOUNTING MANAGER
Kelly Marshall

OPERATIONS MANAGER & MAKER SHED
Rob Bullington

LOGISTICS COORDINATOR
Phil Muelrath

PUBLISHED BY

MAKE COMMUNITY, LLC
Dale Dougherty

Copyright © 2024
Make Community, LLC. All rights reserved. Reproduction without permission is prohibited.
Printed in the U.S. by Schumann Printers, Inc.

Comments may be sent to:
editor@makezine.com

Visit us online:
make.co

Follow us:
X @make @makerfaire @makershed
makemagazine
makemagazine
makemagazine
makemagazine

Manage your account online, including change of address: makezine.com/account
For telephone service call 847-559-7395 between the hours of 8am and 4:30pm CST.
Fax: 847-564-9453.
Email: make@omeda.com

Make:
Community

Support for the publication of *Make:* magazine is made possible in part by the members of Make: Community. Join us at make.co.

CONTRIBUTORS

In honor of spring cleaning, which of your tools or workspaces is in need of a deep clean?

David Battino
Folsom, California
(Rising to the Acacia)
Ha! My desk is buried under four toy robots, three rubbery-mouthed coin banks, two plastic pigs, and an electronic chicken monster.

Claire Danielle Cassidy
Portland, Oregon
(DIY Graffiti Projector)
My 30W CO₂ laser cutter. I hacked it to be quieter and less fumey, but it needs an annual deep clean.

Joshua Pearce
London, Ontario, Canada
(Walk This Way)
Our lab — we have been saving tons of waste plastic with the anticipation of putting a massive, open source hot press into operation this spring.

Issue No. 88, Spring 2024. *Make:* (ISSN 1556-2336) is published quarterly by Make Community, LLC, in the months of February, May, Aug, and Nov. Make: Community is located at 150 Todd Road, Suite 100, Santa Rosa, CA 95407. SUBSCRIPTIONS: Send all subscription requests to *Make:*, P.O. Box 566, Lincolnshire, IL 60069 or subscribe online at makezine.com/subscribe or via phone at (866) 289-8847 (U.S. and Canada); all other countries call (818) 487-2037. Subscriptions are available for $34.99 for 1 year (4 issues) in the United States; in Canada: $43.99 USD; all other countries: $49.99 USD. Periodicals Postage Paid at San Francisco, CA, and at additional mailing offices. POSTMASTER: Send address changes to *Make:*, P.O. Box 566, Lincolnshire, IL 60069. Canada Post Publications Mail Agreement Number 41129568.

PRINTED WITH
SOY INK

FROM THE EDITOR'S DESK

BETTER THAN A DEGREE

I don't normally share praise but this reader feedback made my day. You took a chance on my book, Dale, and I will never forget that. You made it possible. Thank you. *—Charles Platt, author of* Make: Electronics *and many more* Make: *books*

Starting *Make: More Electronics*. I enjoyed your previous work. Your writing style is straight to the point, easy to understand but not dumbed down. Made me a better engineer. Probably as valuable if not more than my electronics degree. *—Joseph Boyle*

FULL CIRCLE

Thank you so much for inviting me to contribute to *Make:* mag. It has been a dream I didn't know I had. I first got into electronics after buying the *Make: Electronics* book, and now I've contributed to it in some small part. How things go full circle. *—Owen McAteer, author of "Flip-Dot Animation,"* Make: *Volume 85, page 68*

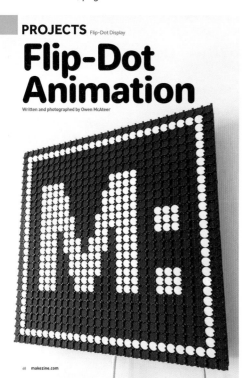

PROJECTS Flip-Dot Display

Flip-Dot Animation

Written and photographed by Owen McAteer

68 makezine.com

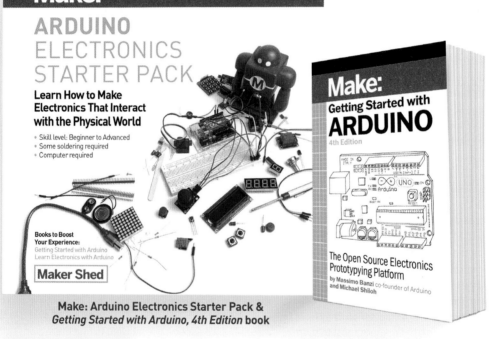

Indie Makers

by Dale Dougherty, President of Make: Community

In 2008, David Pescovitz wrote one of my favorite articles for *Make:* (Volume 13), a profile about John Gaughan, who is a legendary maker of illusions — a magician's magician, if you will. He's not famous but the magicians like David Copperfield and David Blaine who hire him are.

Back then, I was lucky to join David on a visit to the Los Angeles warehouse where Gaughan kept his workshop and museum of the history of magic. "Every surface, every shelf, in these cramped quarters is packed with apparatuses and ephemera that once delighted audiences," David wrote. There was nothing corporate or academic about where John worked. Walking around, listening to his stories and looking at his collections, I was enchanted. The work he was doing was something he'd wanted to do since he was a young kid, intrigued by magic. He just loved doing it. He's had plenty of work, so he chooses to work on the projects that interest him the most.

I ran into David Pescovitz again at Maker Faire Bay Area in October. David was at Mare Island with his son, Lux, who was born the year of the first Maker Faire in San Mateo and is now 17. Lux said what he liked most about Maker Faire were the "indie makers." I asked him what he meant. He said that Maker Faire wasn't corporate and he liked that makers were doing their own thing, exploring whatever they found interesting. For indie makers, it's personal. They have the freedom to do what matters most to them.

Lux's "indie" comment crystallized my thinking about what was different in this edition of Maker Faire Bay Area, why it felt good to have moved outside of Silicon Valley, and how to understand the value of makers and their place in industry and society. Makers play an important role in many creative and technical industries, but it's hard to explain how they influence those industries.

Makers are like indie filmmakers in the film industry. Independent films are made outside of the Hollywood studios and they exist because a person wanted to see his or her film made. Same is true of indie music. In industry, work revolves around those who fund it. If your funder is a corporation, foundation, or government, your work will need to align with their agenda, their priorities. To be indie is to work outside that context and find another way to do what you want.

Being an outsider is not the point. The indie community provides a more open and less formal way to develop new ideas and recognize talent. The community becomes a loosely connected ecosystem or movement, as it is with makers. They share a lot in common but each is different.

There's not a high wall separating what makers do and what professionals do. The skillsets and creative processes are similar. You might see a young maker who created amazing projects in school go on to become productive in the right kind of company. You might find that professionals seek opportunities outside of work to explore their creative interests independently. Makers might go back and forth between a day job in industry and a passion project at night.

The key for makers is having the independence to choose their own projects and bet on their own ideas. Some succeed, some fail, and some are never finished, but the idea was important enough for them to devote time and energy into developing it. They get to choose the work that interests them most. Many people don't get to do that with their lives but makers do, just like John Gaughan. The work is rewarding on its own terms but it's especially gratifying to find there's an audience, a community, who delight in what you do — as you will find at Maker Faire. ◕

Lux Sparks-Pescovitz, David Pescovitz, and Dale Dougherty at Maker Faire Bay Area.

Dale Dougherty

2024
OPEN HARDWARE SUMMIT

The Open Hardware Summit is the world's first comprehensive conference on the Open Source Hardware movement.

Speakers include world renowned leaders from industry, academia, the arts and maker community. Talks cover a wide range of subjects from electronics, mechanics to related fields such as digital fabrication, fashion technology, science hardware devices, and IP law.

KEYNOTE SPEAKER
DANIELLE BOYER

HOSTED BY
CONCORDIA UNIVERSITY & LESPACEMAKER

@OSHWASSOCIATION

KEEP UP WITH #OHS2024 AT 2024.OSHWA.ORG

MADE ON EARTH

Amazing builds from around the globe

Know a project that would be perfect for Made on Earth?
Let us know: *editor@makezine.com*

WALTER, MONICA, CLYDE & TALULAH

DILLUHLSIONAL.COM/AIRFLOAT

In 2012, **Walter Dill** found a 1956 Airflow Land Yacht in Atascadero, California, gave it the name *Talulah*, and set to work renovating it. He used his box truck *Clyde* to haul it to Whidbey Island in Washington and then to New Albany, Indiana, where it currently resides.

"The first task upon returning home was removing the skin," said Walter. "Much of the wood frame was rotted. I removed all the existing wall structure and replaced it with structural panels called Sing Core panels." The electrical and plumbing were ripped out, as well as appliances. "We had to start from scratch," said Walter.

The original trailer was 30 feet long and 8 feet wide, shaped like a "canned ham" with portholes and distinctive corrugated aluminum siding. "All the corrugated siding was removed, treated in huge dip tanks, then painted with Kynar, an industrial aluminum finish typically used on large commercial structures," said Walter. The siding was trimmed to fit the new curve design.

The "retro futuristic" look evolved over time. "**Monica Uhl** came into the project after the basic shape was well underway," said Walter. "She had significant input and involvement in most of the details after that point." The interior aesthetic depended on found elements such as the 1920s cast bronze light fixtures, and incorporates intricate handmade designs using maple, walnut, madrone, sapele, wenge, bamboo, and painted plywood with brass and steel accents. The parquet floor alone includes 10,000 pieces.

"Virtually all the work on *Talulah* was done in and around *Clyde*," said Walter. "I would store all the tools — table saw, bandsaw, chop saw, etc. in the bins around *Clyde*. Our ShopBot Desktop CNC router resides in *Clyde*. A nice little vacuum system resides in one of the bins."

The typical reaction to *Talulah*? "A jaw drop," he said. "People say they could never spend this much time on such a project." —*Dale Dougherty*

Don Wodjenski, Walter Dill, Monica Uhl

I hope... (2021) on display at the König Galerie, Berlin, Germany.

THE TIES THAT BIND CHIHARU-SHIOTA.COM

Have you ever admired the glistening threads of a spider's web in the morning dew and wondered what a super-sized web might look like? How would its many patterns, angles, and connections translate into reality? **Chiharu Shiota's** signature art installations come close, utilizing many interconnected threads to explore abstract concepts of meaning and memory.

Born in Japan in 1972, Shiota has worked as an artist for the last 26 years in Berlin. She doesn't often plan her pieces in advance, preferring to use the existing floor space as she sees fit. A single installation usually uses 250 to 300 kilometers of her chosen material — almost 200 miles worth of black or red thread, rope, string, and wool. For Shiota, the art process is not so much about the idea as it is about the experience. "The entrance is the most important part," she says. "When I set up at the exhibition space, I create the path for the visitors first. Then, we start with the net on the ceiling and on the back of the wall, and then we connect it with the floor. I make art in the space, like painting on a canvas."

Shiota also likes to incorporate hundreds or even thousands of everyday household objects into the mass of threads. Keys, stones, papers, glasses, shoes, clothes, boats, chairs, and more hang suspended in mid-air to illustrate the connected nature of even the most mundane objects in the human experience.

The size and scale of Shiota's installations are impressive, but focusing solely on those aspects misses the most important part — the memories implicit in the empty spaces. "When I started weaving in an empty apartment for one exhibition, I asked the curator who lived there before, and they said an old woman. I thought about her life, how she used the apartment's space, why she stayed there, and the traces she left behind — even though I'd never met her. Especially when people die, the memory stays."

Many of the ordinary items she includes were left behind at flea markets, and she chooses them because they were so important to their owner in life but have lost all meaning in death. "The material is not so important," Shiota states, "but its meaning is much more interesting. Contemporary art has no concrete answer, and it's important that everyone can feel differently."
—*Marshall Piros*

Sunhi Mang / © 2024 Artists Rights Society (ARS), New York / VG Bild-Kunst, Bonn

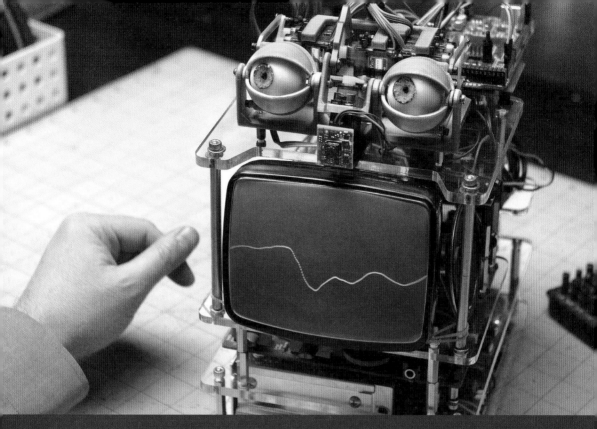

BRIDGING THE UNCANNY VALLEY THOMASBURNS.NET

Thomas Burns grew up in a world where robots like *Short Circuit*'s Johnny 5, *Doctor Who*'s K9, and *Star Wars*' R2-D2 set the public's expectation for what human-robot interaction might look like in the near future. He also spent a good chunk of his life in the country of Georgia, tearing down and cannibalizing old Soviet tech, as exemplified by the benchtop power supply he fashioned from an old TV. He came to electronics relatively late in life, after a couple of decades in cinematography and visual storytelling. Like so many of us, it was an Arduino kit that opened a door "to this brand-new creative universe" that kept Thomas up at night as he marveled at the possibilities afforded by this new skill.

So how did he go from budding electrical engineer to creator of an enchanting Alexa-powered robot with over 3.5 million views on YouTube?

First and foremost, Thomas sought to fill the disparity between today's technology and what he'd imagined as a child. He saw a Muppet-like computerized creature as far more approachable than humanoid robots — which somehow become even less appealing as their "realism"

increases. In order to suspend disbelief and make you forget you're talking to a machine, Thomas wanted his Alexatron to create natural-feeling eye contact — so he borrowed a design from fellow YouTuber Will Cogley, using six servos to move the eyeballs and lids in an engaging manner — while intentionally retaining some endearing "ricketiness" and embracing servo whines. To create an appealing analog representation of the voice, Thomas leveraged an old hack that generates an oscilloscope-like signal on a small CRT, evoking nostalgia for a time before screens were flat and digital.

In addition to the titular Alexa, an Arduino Mega ties things together, itself woken up by a repurposed LED pad on the Echo Dot. An Adafruit servo driver makes it move, while TDA7297-based amps boost the audio and visual waveform. Finally, a Useful Sensors Person Sensor tells it where to point those peepers. The result is an uncannily engaging blend of nostalgia and new tech that you can plop in front of a 5-year-old with no explanation and marvel at their effortless interaction. *—David J. Groom*

Project Empire lights up the Esplanade at dusk along the Napa River.

Maker Faire Bay Area Brings Makers Together Again

Written by Dale Dougherty

THE MAGIC IS BACK BY THE BAY

DALE DOUGHERTY is the founder and publisher of *Make:* and creator of Maker Faire.

EepyBird's Diet Coke & Mentos show returned to Center Stage.

Cupcake Car from Acme Muffineering.

"Fantastic," I heard someone call out, on the sixth and final day of Maker Faire Bay Area, as makers were packing up, ready to take their creations back home. Keith Johnson was climbing on Jon Sarriugarte's *Project Empire*, the enormous, bright green space vehicle, and as I got closer, Keith hollered out: "Maker Faire was fantastic." I was happy to hear that. I was happy to see the relaunch of Maker Faire Bay Area at Mare Island and hear that everybody loved the new waterfront venue. I was happy that so many makers returned who had been at Maker Faire before as well as a whole group of first-time makers.

Keith Johnson and Merrilee Proffitt, who brought their Cupcake Cars to earlier Maker Faires in San Mateo, brought *Electric Wrecker*, a large, four-wheel vehicle they had built. At the first Maker Faire, Jon and his wife Kyrsten Mate brought the *SS Alpha Fox*, a NASA-inspired rover, and in later years *The Golden Mean*, a giant snail. Now they brought *Project Empire*, which had a crew of 25 people working on it for eight years (see *Make:* Volume 87, page 128). At the first Maker Faire, Kyrsten was pregnant with their daughter Zolie — who's now a teenager and performed as a crew member of *Project Empire* alongside Keith and Merrilee's daughter.

After a four-year hiatus, Maker Faire Bay Area returned in October 2023 for two weekends along the waterfront at a historic naval shipyard, Mare Island in Vallejo, California.

I didn't really know what to expect by bringing back Maker Faire Bay Area. It was something of a risk. Would enough makers show up with projects? Would enough people come out — those who had been to Maker Faire in the past and those who didn't quite know what to expect?

Would it feel as wide open and full of creativity and brilliance as it once had? Maker Faire was coming back to a world that was very different from when it first launched.

Perhaps I need not have worried at all.

Makers at the Heart

The magic of Maker Faire returned along with a fascinating collection of makers and eager attendees who were glad to see it back. You could feel it in the air. You could see it in the smiles on faces. People on buses coming from the parking lot to the entrance were buzzing with anticipation, and they were buzzing again with excitement for what they had experienced when they boarded the buses to go back to their cars.

The crowd that gathered for EepyBird's Diet Coke & Mentos show, a popular spectacle in past years, was buzzing too, like it was their first time — and it might have been because the crowd was so young. So much of Maker Faire is encountering the unexpected, like the

Dale Dougherty, Mark Madeo, Ben Didier

Irma Harris dazzled crowds as robotic humanoid bird *Valkylrma*.

Taylor Waddell demonstrates UC Berkeley's Computed Axial Lithography (CAL) instant resin 3D printer.

Stephen Jacobsen, Keith Hammond, Ben Didier, Mark Madeo, Dale Dougherty

robotic humanoid bird *Valkylrma* by Irma Harris or the new Computed Axial Lithography (CAL) instant resin 3D printer demonstrated by UC Berkeley researcher Taylor Waddell (see page 38 of this issue).

Each day we had a parade — a rather impromptu event that makers joined with anything that moved — led by Russell the Electric Giraffe, followed by colorful, ridable electric muffins, hand-held kite puppets, a shark-shaped motorcycle, the pedal-powered *Trashlantis* kinetic sculpture, the rumbling Holy Bike, a variety of mobile creations from Obtainium Works such as a magic lamp spouting fire with a genie inside, a marching band in red, white, and black uniforms, a red robotic cart with big speakers, and finally on the last day, even Kinetic Steam's giant steam engine rumbled down the wide Esplanade. People stood on both sides of the parade, lining up as you would on a street to watch, take photos, and wave. It was glorious.

This Maker Faire was something that only this community can do, and only at Maker Faire could you experience it for yourself. This new edition might have been smaller than previous editions of Maker Faire Bay Area, except maybe the very first one, but it had all the essentials, such as learn-to-solder and many more hands-on activities. It had more than enough heart and soul to fill a vacant shipyard. The energy and enthusiasm felt new — it was not something

we've been feeling during Covid, when many of us were isolated, or even since then. Over the last 10 years, our culture has changed and become darker and more dire, but Maker Faire was like a string of LEDs offering hope, shining on the many creative interests of so many people and suggesting what is possible. It was great being together again.

Who knows what impact Maker Faire has on people's lives? I do hear from attendees — and from makers returning year after year — what it means to them. The kid who now works as an engineer and was first inspired by Maker Faire to tinker. The now-grown son of a maker who has his own custom vinyl record business because he figured out how to make records in small batches. The students from Berbawy Makerspace at Irvington High School who managed the Nerdy Derby activity for six days, one of whom told me she had no idea what Maker Faire was when she came but this had been the best experience of her life.

Sure, there were plenty of robots and rockets, drones and LEDs, welding, soldering, knitting, glass blowing, and metal working at Maker Faire. There was a Dark Room, which featured MakeFashion Collective putting on Hack the Runway shows the first weekend. Erin St. Blaine walked down the runway in an LED dress, passing underneath her glowing jellyfish sculptures. Craig Newswanger's *RayLights*, mounted on a wall, generated mandala-like color

Russell the Electric Giraffe cruises around Mare Island Naval Shipyard.

Pedal-powered kinetic sculpture *Trashlantis* rode in the parade.

Keith Young rides his *Holy Bike*.

The drivable *Genie Lamp* from Obtainium Works joined in the parade.

Students from Berbawy Makers ran the Nerdy Derby races.

School kids gathered around a quadraped bug robot.

Erin St. Blaine in MakeFashion show with her *Jellyfish Grotto* above.

Craig Newswanger's *RayLights* reacted to sound in the Dark Room.

Celestial Mechanica by Jessika Welz dominated the Foundry space.

Sepia Lux lit up the dark room during the second weekend.

Adam Savage gives his Sunday sermon.

patterns based on ambient sounds. The following weekend, there was a host of light-up creations including *Sepia Lux* — the animatronic cuttlefish art car. In the cavernous Foundry, an angular three-story building at the opposite end of the Esplanade, was *Celestial Mechanica*, a large-scale, motorized orrery.

By Making Things, We Create Ourselves

On the first Sunday, Adam Savage gave his Sunday sermon, a tradition resumed. He arrived atop Russell the electric giraffe, coming up behind the crowd who were already sitting in rows waiting for him. They all turned to see his familiar face again. Adam began by saying that he had filmed dozens and dozens of episodes of *Mythbusters* on Mare Island.

Adam said that he wanted to talk about "making as self-improvement." What he wanted to say would be "weirdly, deeply personal" so he advised his listeners to "strap in." After a short pause to look at his notes, he said: "The first thing I want you to know is that I was a lonely kid. I didn't make friends easily and I didn't know how to act around other kids. I was really lonely. And, of course, it doesn't make me unique or special."

Seeking a way out, he made things — endless, elaborate things with Lego, with papercraft — as a way of building a castle wall between his conscious self and his loneliness:

Mark Madeo, Stephen Jacobsen, Ben Didier

Flames erupt out of *Serenity* from Flaming Lotus Girls.

"The amazing thing about being a person is that every one of our coping mechanisms carries with it not just the tragedy that necessitated it but also a superpower. Small me, self-soothed by concentrating on tiny things carefully, specifically paying minute attention to endless details and getting lost in them and the stories that were created around them. I was using that practice to drown out the fact that I didn't know how to have friends but at the same time, that practice, that iterative process of ideation, creation, and play was building something vitally important in me. My very self was forming around the thoughts and ideas that I couldn't get out of my head. As my creations improved, so did my confidence and self-esteem. ...

"This is for me the central reason that I make. It's because when I make, and I pay attention to the things I'm making, I learn important things about myself, and about being a better person. That's why, I believe in the end, we all make."

Golden Possibilities is a 10-foot-tall welded metal palomino horse from artist Pierre Riche.

That lonely child is still with Adam every day, he said, no matter how busy he is, no matter how popular he is. He's become a better person who makes things, who loses himself for hours at a workbench, who is able to connect to people intimately in large crowds and tell his story. It's a story most makers can identify with. His struggle, our struggle continues, and we keep going; we keep making and becoming better humans.

With large rusted cranes looming above our heads, we knew that big things had been built at the Mare Island naval shipyard. With all of the makers gathered here, Maker Faire Bay Area was a big thing too. All who participated in Maker Faire were happy to feel the creative vibe again, realizing how necessary it is to make things for yourself and with others. On the *Make:* YouTube channel, user @rikaika4178 commented on Adam's Sunday sermon: "Maker Faire motivates me like nothing else to learn a new skill, tackle a new task, and to find a way to give life to that little spark that's been rattling around in my imagination." That's the magic. That's fantastic. ●

Hack Club's Chris Walker sailed the aquatic bounce house *Castle Bravo* along the Napa River.

There were many hands-on activities to explore.

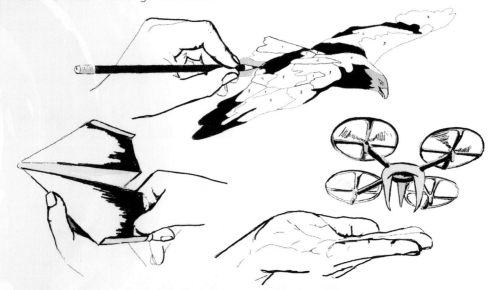

3 MODES OF MAKING

Written, illustrated,
and photographed
by Matt Zigler

Try the trifecta of teaching maker skills:
imitation, modification, and innovation

Making things can be complex and challenging work. There are times when we act as researchers, learning some new tool or technique. We scour the internet for tutorials, talk to friends, or look at back issues of *Make:* to help us gain the knowledge we need. Whether learning a new woodworking tool, microcontroller, or sewing machine, success requires patience, attention to detail, and a willingness to fail and try again. Imitating the work of others to learn new skills is one of the central tasks of a maker.

But that's just the beginning. Once we have a level of confidence with our tools and procedures, we have to figure out how to make something that actually suits our needs in real-world situations. We have to pull out measuring tools to make sure our piece of furniture fits in the space available. We have to know enough about the properties of

our materials, physical or digital, to understand what we can and cannot accomplish with them. These are not only technical challenges, but also mental challenges that test our ability to plan, predict, and problem-solve. Being able to take the world as it is and modify it to our purposes is how makers take ownership of our surroundings.

And as if that weren't enough, the process of generating ideas offers a whole other level of complexity. If we get a flash of inspiration, how do we go about pursuing it? If we're struggling to come up with an idea, how can we trigger our creative juices in order to create something new and interesting? What should the first prototype be and what are the biggest challenges standing in the way of a successful build? The exploration and experimentation involved in making something innovative can be as intimidating as it can be exciting.

Traditional education does its best to prepare us to do high-level algebra and calculus, as if we will be doing complex mathematical modeling on a daily basis. The typical curriculum doesn't work quite as hard at helping students develop the many skills I mentioned earlier. These skills are useful for makers but also for anyone who does interesting work for a living. A lawyer doesn't just need to research legal precedent, they also need to understand their clients' needs, predict the biggest challenges in a case, and craft a persuasive explanation of the facts. A shop owner needs to arrange displays so customers want to come back and must always be looking for new ideas and products to grow their business. As a teacher in a school makerspace, I see it as my duty to not just teach students technical skills, but also these other, less tangible skills.

Paint by numbers is a classic imitation project.

Imitation, Modification, Innovation

For the past several years as a maker educator, I have been using the concepts of ***imitation, modification, and innovation*** as the structure for how I think about the curriculum I teach, how I assess students, and most importantly, what I want them to get from their experiences in the makerspace. Expertise and mastery over the skills involved in these three modes of making are universally applicable in any pursuit that involves creative thinking and problem-solving.

It's important to think of these three modes as equally essential and connected. There is often a tendency to imagine that imitation is only valuable to pursue modification, which leads us to innovation — that imitation is in some way a lower-level skill. Many times, I have heard maker educators dismiss "kit projects" as lesser than other projects, but this shows a focus on product over process. When we focus on the process of imitation, we can look at STEM kits, tutorials, and pre-packaged projects and evaluate them on how well students can practice the skills of imitation.

"Imitation: Working from a set of steps to build foundational skills, basic knowledge, and hands-on experience."

Imitation, when done well, is about improving technique and concrete skill acquisition. It is a way of learning from what has come before and can be done accidentally or intentionally.

Modification can also get a bad reputation in that it is considered a surface-level activity at best and, at worst, theft or copyright infringement. Modification can range from engraving your name onto a water bottle to altering existing code or taking a design intended for one purpose and applying it to another. It is taking something that already exists and modifying it in a way that suits you or another better. Taking some level of ownership of the object and design is a crucial factor. Open-source design and Creative Commons licensing are both based on the idea that we may take designs and tweak, alter, and improve them while recognizing the chain of creators that led up to our contribution.

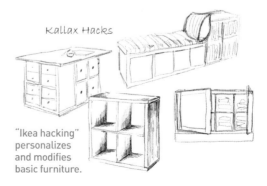

Kallax Hacks

"Ikea hacking" personalizes and modifies basic furniture.

"Modification: Remixing existing ideas and objects with tools, materials, and techniques with which you have become familiar."

The result of a modification project doesn't significantly change the function or application of the original. The engraved water bottle is still a water bottle, and the edited code doesn't serve a radically different function; they are both improvements over the original while maintaining the best qualities of the original object. As with imitation, many skills can be learned through modification, such as understanding constraints, creating a plan, and working with the physical properties of materials.

The stated goal of many makerspaces is innovation, but that's difficult to define. Often it is easier to point at it than to explain what it is.

"Innovation: Combining previous experiences and skills to create something 'new' and bringing those ideas into the world."

I put the word *new* in quotation marks because there are no genuinely new ideas, in the sense that they come out of nowhere. New ideas are significant steps taken beyond existing ideas, pushing them into uncharted territory.

Innovation happens through exploratory processes such as prototyping and iteration. Innovators are feeling their way toward a stated goal, relying on skills and experiences they already possess, but also figuring out things on the fly as needed.

An innovative project often begins as a form of modification but goes a little further. As students move into new territory, they'll often need to rely

Asking how many uses you can come up with for a paperclip is a classic test of divergent thinking.

on imitation to pick up the skills required to make that next step. Innovation cannot be separated from modification and imitation because it relies on them to reach the goals of the innovator.

This method of making can be the most daunting for teachers in the makerspace. Students will inevitably come up with ideas and goals beyond the teacher's skills, but this is how students learn creative problem-solving, and the teacher is there to help guide them through the process, even though they may not be able to help them with the product.

Long ago, I gave up on being the "sage on the stage," standing at the front of the room and passing valuable knowledge to my students. Every year my students want to pursue projects that are beyond my knowledge and abilities. I let them know ahead of time that they will need to pursue their own sources of information and knowledge. I can help them ask questions, seek sources, and think through their problems, but I can't tell them how to do it. Rather than feeling unqualified or powerless, I feel proud that they are willing to risk an incomplete or failed project.

None of the skills involved in these three modes of making are new and groundbreaking. Attention to detail, finding personal meaning in projects, and looking at the big picture are all things we do in various ways at a very young age. The ability to do them with purpose and to learn from the intentional application of these skills is what students can get from a maker curriculum based on imitation, modification, and innovation.

Maker Skills in Traditional Courses

At my school the makerspace works more like a library than a tech class. While we do offer a variety of technology- and design-focused courses, the majority of students use the space for projects based on topics they're studying in their traditional courses. This not only broadens the exposure of students to the makerspace, it reinforces the importance of applying these methods in all areas of life. Students who participate in these projects are often eager to take a more in-depth class.

Students in world history study ancient civilizations, learning about artifacts, societies, and cultures. What if those cultures were still around today, designing objects for our contemporary world? What sort of symbols might they embed into today's products? If writing and communication were of the highest importance, how might the communication devices we use be designed? If nature was commonly represented, how might that find its way into today's consumer products? This project (Figure A) requires students to make connections between the world they know and the world they are studying, involving both modification and some innovation.

Maker projects in science classes often involve ways to create instruments or present factual information in new formats. These projects tend towards deciding on a plan and enacting it step by step, involving a lot of imitation and modification. One such project in an astronomy class explores the relative distances of stars within a single constellation. Constellations only appear to have their shape from our vantage point on Earth. In reality each star can be hundreds or thousands of light-years closer or farther away from us. By using layers of engraved acrylic, and an editable 3D-printed base, students create physical models of a chosen constellation, representing the three-dimensional reality of the seemingly flat pattern. When viewed from any angle other than straight on, the engraved stars shift into new arrangements (Figure B).

World language classes often require students to speak about themselves in order to practice conversational or presentational language. This Spanish project uses the form of a traditional textile from Panama, called a *mola*. A mola uses lots of high-contrast, bright colors, and rich symbols and imagery. The layers are cut away to create intricate patterns and designs. In this project students create a design with symbolism that represents themselves; their experiences, passions, and motivations. Using layers of felt or other fabric, the designs are cut with a laser cutter and assembled by hand using glue. Students will later use this as a display when presenting to the class (Figure C). This project is all about modification, taking rectangles of fabric and shaping them to create a personalized object.

The next time you're working on a project, think about what mode of making you are using at the moment: imitation, modification, or innovation.

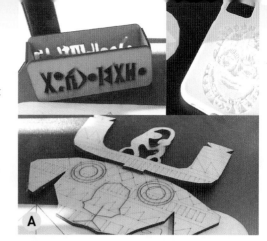

A

From top left: Ancient Sabaean tealight holder, cellphone case with design from Benin, contemporary dog mask inspired by the Mossi people of West Africa.

B

C

What is your brain doing when you're in that mode? Which modes feel most natural? Which are you most uncertain with? By understanding our process, and pushing ourselves a little bit more with each new project, we can continue to expand our mental toolbox as makers to accomplish whatever we set out to make. ◐

This article is adapted from the new book *3 Modes of Making; Designing Purposeful Projects to Teach Maker Skills*, now available at the Maker Shed (makershed.com) and fine booksellers.

MATT ZIGLER is an artist, teacher, and maker. He is the Bullis Innovation and Technology Lab coordinator at Bullis School in Potomac, Maryland.

MICRO RECYCLERS

SAVE THE WORLD AND HAVE FUN MAKING STUFF WITH THE **PRECIOUS PLASTIC** COMMUNITY

WRITTEN BY NICK SCHICK WITH ANDREA SCHICK

David Schick, Andrea Schick, Precious Plastic

A few years ago in 2020, my dad showed me this great YouTube video called "Precious Plastic — The Story Behind" (youtu.be/EPA2l1bi2pQ). We were inspired by the Dutch grad student who created the video, Dave Hakkens, after hearing him talk about the plastics problem. He told us how there was plastic waste all over the world. Globally only about 10% of the plastic we put in the recycle bin actually gets reused. The vast majority of it goes into our landfills, streets,

oceans, and even into our bodies — according to a study commissioned by the World Wildlife Fund, the average American consumes up to a credit card's worth of plastic every week. And, CO_2 levels everywhere in the world are rising, accelerating climate change.

My friends and I wanted to do anything we could to reduce global warming, and we knew reducing plastic waste was one way to contribute. Hakkens talked about simple ways to repurpose plastic as part of an international community — hundreds of groups of makers around the world who were sharing ideas online and even equipment, learning ways to innovate and repurpose plastic waste.

UPCYCLING NOOBS

I got excited about the idea of repurposing and shared it with friends at school. Soon, six of us began working on projects in my family's shop and we started Peninsula Precious Plastics. Tomas Kolecar of Plastmakers in the Czech Republic sent us molds for making clocks and clipboards. Andrés Garzón (El Tornillo) sent molds for flowerpots from Bogota, Colombia, and we got a Version 1 shredder from Papa Diouf at Precious Plastic New York. Next, we purchased an industrial granulator from Maselli & Sons Hardware in Petaluma, California, which has to have the greatest hardware salvage yard ever!

We learned that to recycle and repurpose properly, it's best to remove labels from the plastic items, then throw them in the dishwasher. Afterward, we sort them by different types, i.e. HDPE and polypropylene (resin types 2 and 5, indicated by the numbers on the bottom). We learned which ones are best for making clipboards, clocks, flowerpots, and keychain fobs, and started documenting color combinations and temperatures.

It's really fun to see the swirls of colors, like magenta, purple, and green, and come up with monikers for them like Purple Haze, Mardi Gras, and Haight Street. We also had beautiful blues, greys, yellows, and greens that looked like the ocean, and white clipboards with small splotches of color we called Confetti. We enjoyed coming up with color combos, and when I showed the clipboards to some of my friends' parents who

We affectionately nicknamed our shredder "Crunch Bird."

Shredded plastic ready for projects.

Nick Schick and father David Schick show off clocks molded from recycled plastic.

The global Precious Plastic community shares open designs for building this plastic shredder, extruder, and sheet press, among other machines.

Peninsula Precious Plastics workshop with students from Aragon High School, San Mateo, and Carlmont High School, Belmont, California.

Precious Plastic's map of nearly 1,000 recyclers and makers, machine shops, collection points, and communities.

were doctors, they said they'd like to use them at their hospitals.

Dr. Christina Lee told us, "Not only are you saving the environment with your efforts, the clipboards are beautifully and wonderfully made." She wound up becoming the sponsor for my Eagle Scout project, and my Boy Scout Troop 42 got involved, too, making clipboards that I presented to Laguna Honda Hospital for their vaccine clinics. Later, one of my troop leaders,

Dr. Ken Lin, noticed how much clean plastic his hospital disposed of daily that could be reused, which he began giving us. "Once you start looking for wasted plastic," he said, "you see it everywhere." Many teachers also started asking for our clipboards for their schools.

SHARING GLOBALLY AND LOCALLY

We also came up with cool ideas about how to share what we made online. Callie Shawnte,

Flowerpots from recycled plastic.

Mini press and oven for molding.

Sheet press.

Clipboards with [L-R] Oliver Piquet, Ari George, and Delaney Wolfe in San Carlos, California.

A 1-meter square repurposed plastic sheet, still warm off the sheet press

David Schick, Andrea Schick, Precious Plastic

Precious Plastic in Switzerland: Verdan Deliz and Claude-Anne Schumacher (Glitter Geneva), with Nick Schick, David Schick, and Andrea Schick.

who previously worked in communications for Precious Plastic, showed us how she made Instagram reels that were really engaging. Tomas Kolecar mentioned sharing stories daily to create a broader audience for what we were doing. Students, artists, and makers saw our work and reached out to us via social media.

We wound up hearing from some groups internationally and had the privilege of going to their makerspaces and sharing ideas with Glitter Geneva in Switzerland and Plastmakers in Czech Republic. Some of these interactions we shared on YouTube, Facebook, and our Instagram site. Tomas also mentioned it was important to know public policy, as laws regulate plastics and plastic production and repurposing. So we asked local nonprofits, school administrators, and California state legislators to support our movement.

Many of my friends in high school wanted to participate, because we had creative innovation and tech ideas as well as presentation skills. When we applied to colleges we wrote essays about our experience, and my friends and I wound up in amazing universities such as UCLA, USC, University of Chicago, Johns Hopkins, and Georgia Tech. California State Parks also loved our ideas about repurposing and said we were welcome to present at their locations. Local discovery museums, such as San Mateo's CuriOdyssey and San Francisco's Exploratorium, also reached out to us. Friends are now looking into starting Precious Plastic groups on their college campuses.

Other local high schools reached out and asked us to share information, as they wanted to start programs at their schools. We presented at our high school's International Fair, about our international community of makers who share ideas globally. We also received presentation requests from local AP Environmental Science classes, and we were able to showcase our work at school club fairs and at the 2023 Maker Faire Bay Area at Mare Island.

To gather waste plastic, we made bins where we could collect it at schools. My Boy Scout troop families began bringing plastic waste over to our house, and then we created a separate bin there, so people could drop off plastics for our repurposing projects.

CHANGE IS GOOD

We felt good about all of this, because as I said earlier, 80–90% of our plastic waste goes into landfills and oceans, or litters streets around the world. So as long as plastic waste is getting thrown out — even in recycle bins — it is a problem. It's clear we need to slow down our rampant plastic production, but even if we do, plastic won't just go away.

And plastic is an extremely useful material: incredibly malleable, easy to sterilize, waterproof, lasts hundreds of years. We seek to use this material and take the plastic problem into our own hands by making locally and sharing ideas internationally to make a dent in climate change worldwide, one person and one school or community organization at a time. ◉

LEARN MORE:
- Peninsula Precious Plastics: instagram.com/peninsulapreciousplastics
- Precious Plastic Academy: community.preciousplastic.com/academy
- Precious Plastic Map: community.preciousplastic.com/map

NICK SCHICK is a maker, environmentalist, and actor/writer/director who wants to educate himself and others about the planet through video, theater, and hands-on fieldwork. A student at UCLA TFT, he is cofounder of Peninsula Precious Plastics.

ANDREA SCHICK is a writer, editor, and events manager for STEAM nonprofits, and cofounder of Peninsula Precious Plastics. She loves nature, travel, the arts, education, and learning from the creativity, innovation, and artistry of makers everywhere.

DIGIFAB
2024

THE RISE OF COLOR

IN 3D PRINTING

Written by Courtney Blum

TRUE MULTI-MATERIAL MACHINES ARE HERE — AND THEY'RE GOOD!

COURTNEY BLUM aka Filament Stories is a short-form content creator talking with great enthusiasm about all the myriad colors and styles of 3D printing filament and keeping us in touch with the latest trends and sheer joy that a spool of plastic can be when turned into beautiful 3D prints.

I remember the utter dismay I felt when I was looking to buy my first 3D printer. They all seemed to print in just one color. Plastic things were all around us, and they didn't seem to be limited to just one color — so why couldn't 3D printers print in lots of colors, too? It wasn't until I started printing that I began to understand the technical challenges of getting more than one color into a model that was built one G-code line at a time.

For this article, we'll focus on filament-based 3D printing, also known as FDM (fused deposition modeling) or FFF (fused filament fabrication), specifically in the consumer and hobbyist realm. Color printing does exist at the industry level with a variety of other technologies, but none are at a price point that can be easily scaled down to the consumer level.

What have become accessible and prevalent are the familiar printers that can accurately re-create a 3D model by melting filament of various polymers. So our multicolor problem is, how do you stop printing in one color filament and start printing in a second color? What about a third, fourth, or even 10th color?

Solving this problem has been a challenge that has been approached from multiple directions. In rough historical order, they are:

INDEPENDENT DUAL EXTRUDER (IDEX):
The first multicolor hobby-level printers were developed with two separate extruders, each able to print a different filament, enabling two-color models. These machines can print in multiple ways, with some providing the ability to print the same model twice at the same time. IDEX printers are still popular today and are available from multiple manufacturers, for example the **Sovol SV04** printer.

COLOR BLENDING EXTRUDER: Other companies, like **Geeetech**, took a different approach to multicolor printing by bringing two or three filaments into the same hotend (Figure **A**). This color mixing configuration, along with customizations in the slicer, gives the printer

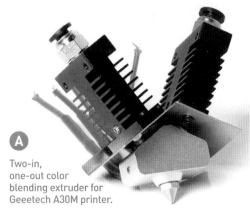

A Two-in, one-out color blending extruder for Geeetech A30M printer.

B

Piles of purge towers from filament swapping multicolor prints.

the capability not only to print in multiple colors but also to blend the colors together. It can print the model with a gradient across colors, or use a percentage blending between the two, or print specific colors in certain locations.

This style of printing has remained more niche, partly due to challenges with nozzle clogs and hardware reliability, and partly because the software isn't mainstream and can be challenging to decipher (I was never able to get it functioning beyond basic gradients). However, many have enjoyed printing multicolor prints this way. Geeetech still sells their two- and three-color printers as well as a new **Mizar M** model that combines both approaches: dual extruders and color blending within one nozzle.

And for those wanting to modify their existing 3D printers, **replacement blended nozzle hotends** have recently become popular, with multiple brands selling hotends with two, three, and four filament inputs.

FILAMENT SWAPPING: To have true multicolor printing, a combination of new hardware and software was needed. In 2017 **Prusa Research** debuted their **Multi-Material Upgrade (MMU)**, an additional filament swapping module for their Prusa MK2 i3 printer that made printing in up to five different colors a possibility. This was coupled with model painting and slicing functionality in PrusaSlicer software to create a G-code file that

could tell the printer to print in one color, retract that color, move to the next color, print, retract, etc. until the model was complete.

Printing in this manner requires an additional step that adds significant waste to the printing process: *purging* filament. Because only one extruder is being used, when one filament is retracted and the next filament loaded, there is still melted plastic in the hotend from the prior color. In order to get to a "clean" color, the prior color needs to be pushed out. This is done via a *purge tower* or *wipe tower*: the extruder lays down lines of filament until it's estimated the old color has been purged out (Figure **B**).

Purges need to be done once per color change per layer of the print, and are deposited over on the side of the build sheet. In the case of four or five colors, the filament used to purge until the next color is extruding cleanly can easily be larger in weight/volume than the model being printed!

FILAMENT SPLICING: Other technologies were also in the works. **Mosaic Manufacturing's Palette** was a solution that could be used with any 3D printer, unlike the MMU which was specific to Prusa's. The Palette 2 (Figure **C**) supported up to four colors of filament, and Mosaic's Canvas web-based tool was used to paint the model and define printing parameters. From there a customized version of the model was sent to the printer, and the Palette was told to begin

Courtney Blum, Carnival and Zigzag flasks by Clockspring3D

preparing filament for the printer.

The Palette managed multicolor printing in an entirely different way: by splicing together the filament into sections as long as the printer would need to print that color on each layer. The Palette worked with any printer because the printer had no idea it was printing in multiple colors. It was receiving a single strand of filament, spliced together by the Palette and fed to it as it printed the model. The concept was elegant, the execution was complex, and the engineers at Mosaic made it all work together.

There was still the purging problem, though. Even though the filament was spliced together for the proper lengths the model would need, those lengths had to be extended so the color change could make it to a pure color (no one wants red bleed into white filament, for example). Which left us with purge towers and a lot of waste. That was just the price we had to pay for changing colors with just one nozzle.

Hold that thought, because there are ways we can reduce that cost.

MANY MULTICOLOR MODIFICATIONS

Palette 2, which had some reliability issues, was replaced in 2021 by the more robust Palette 3, which could print in up to eight colors with more reliability (Figure **D**). In 2023 Prusa Research released their updated MMU3 for Prusa printers, also with more reliability (Figure **E** on the following page).

It is here that you might think you're hearing about a lot of "reliability improvements." As it turns out, printing in multiple colors is hard. Loading one color, unloading it, loading the next color, unloading it, and then repeating that a thousand times is a lot of manipulations with filament. Whether you're just loading or splicing, there are a lot of parts in play to make those pretty multicolor prints.

MINIMIZING PURGES

Carnival and *Zigzag* flasks by Clockspring3D with their purge towers.

So far, we have printer attachments that feed multiple filaments into the extruder, and we have products that splice the filament and send it to the extruder in the order needed for the colored print. Both approaches have one thing in common and that's one extruder. Filament that goes into the hot part of the hotend melts, and the next color that comes in isn't going to come out until the prior color is flushed out. That's what the purge towers accomplish. The purge waste can be significant — I had a 12-gram jar in eight colors that had a 500-gram purge tower. Which is crazy, right?

There are ways to reduce this waste. First, **the amount of purge can be calculated and reduced** to the minimum needed. For white after black or dark red, the amount of purge will likely need to be a good bit to get the dark, pigmented color out. But for dark green to black, the amount needed may be very little. All filament products can have their purge values adjusted, and in many cases, the total waste can be reduced.

Another option is to put the purge material to another use, such as **infill on the object** being printed. Depending on the amount of infill the object has, much of the purge filament will be hidden inside the object being printed.

And lastly, a **second "purge" object** can be printed where color doesn't matter, because the object might be painted, for instance.

Another way to minimize purge material is on a **per-model basis**. For the example where I had 500 grams of purge material for one small model, the purge amount may not be able to be reduced, but the number of models could be increased. Sixteen of that particular model could have fit on the build plate, reducing the per model waste to 31 grams per model.

So, can single-extruder purges be eliminated? No, but they can be minimized a good deal.

Prusa's MMU3 swapper add-on handles up to five filament colors.

Bambu "poops."

More filament swappers soon followed. **3D Chameleon** has a color changer that adds four-color printing to any 3D printer. A recent Kickstarter, **Co Print**, promises fast color 3D printing with up to 20 colors that works on any 3D printer. And there's the **Enraged Rabbit Carrot Feeder (ERCF)** for Voron printers.

All of these are modifications that are added to printers. But none was designed as a true multiple color 3D printer. That is, until **Bambu Lab** came on the scene via Kickstarter and took the world of multicolor printing by storm with their printers that poop.

BAMBU LAB GETS THERE FIRST

Bambu Lab came out of nowhere in 2022 with their **X1 Carbon** printer and its **AMS (Automatic Material System)** that had four-color printing with a filament swapping hub that increased the number of colors up to 16.

The way the filament is managed by these new printers involves a lot of loading filament in through a long series of Bowden tubes, printing that color, unloading it, loading the next color and then, instead of printing a purge tower, the printer makes a little squiggle of filament that it then ejects out the back of the printer (Figure ). It is unlikely Bambu Lab intended the term to be *poop*, but once it was said, the term stuck.

And here's the thing. The reliability and print quality coming from these printers, started by a team of ex-DJI executives, was unparalleled.

They successfully fulfilled their Kickstarter pledges, and sales continued to go up based on print results and user endorsements. Many of the sales were from existing customers buying a second or third printer. Since then, Bambu Lab has released additional printers such as the new **A1** (reviewed on page 125 of this issue) and **A1 Mini** (Figure G), all with multicolor printing capability.

Did this solve the multicolor waste problem? Nope. It only changed the way the waste was created, from purge blocks to piles of poop. Just as in the other technologies, tuning and other options can reduce the waste or redirect it to other areas, but the waste is still there.

A TOOL CHANGE WILL DO YOU GOOD

Does this mean we're stuck with lots of waste for multicolor printing? Not necessarily. In all the technologies above that give us more than two colors, we've had just a single extruder. But there are also *tool changer* 3D printers which have separate extruders for each color or material. E3D sold one in 2018, a proof of technology that is no longer in production, but the new **Prusa XL** from **Prusa Research** (reviewed on page 124) has up to five toolheads that can do multicolor printing with very little waste (Figure H).

Because each extruder maintains a single color, there's no need to purge between color changes; only a tiny *priming* bit of filament is extruded to ensure the filament is flowing

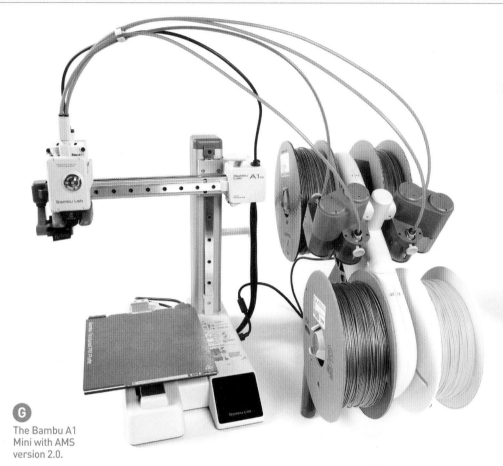

G The Bambu A1 Mini with AMS version 2.0.

normally. Models printed in five colors create only a few grams of filament waste in their little priming tower, vastly less than single-extruder machines.

There is a different price to pay in this case though, and that's cost — about $3,500. Five extruders are more expensive than one. However, that cost may become competitive with single-extruder solutions in the future, with other color-changing machines bringing down prices. One new tool changer currently on Kickstarter is the **Proforge 4** from **Makertech 3D**, which has four toolheads for about $2,000.

And that brings us to our last and latest multicolor printing product that has only just been shown at Formnext 2023 in November: **AnkerMake's V6 Color Engine**. This innovative-looking upgrade for their M5 printers has six little extruder/hotends built into one big print head, where each hotend rotates into place for fast color

H The new Prusa XL can have five independent print heads.

changes without the need for purging the prior color. Is this the next advancement in fast, low-waste multicolor printing we've all been waiting for? We hope to test it soon and find out! ●

Prusa Reasearch, Courtney Blum, *Planetary Phone Stand by Clockspring3D*

TO MULTI OR NOT TO MULTI?

WE TEST THE QUESTION

Written and photographed by Caleb Kraft

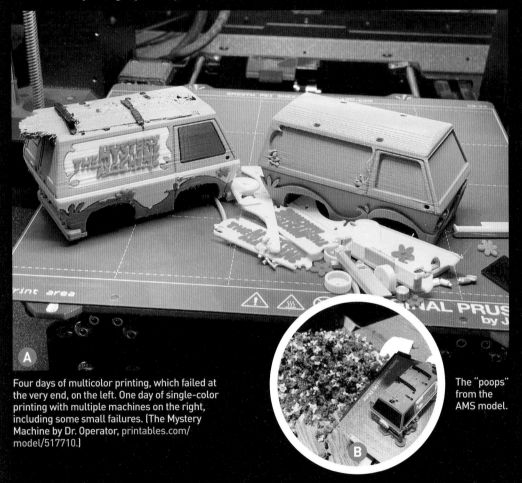

A Four days of multicolor printing, which failed at the very end, on the left. One day of single-color printing with multiple machines on the right, including some small failures. (The Mystery Machine by Dr. Operator, printables.com/model/517710.)

B The "poops" from the AMS model.

CALEB KRAFT is senior editor for *Make:* and has been 3D printing for over a decade. He still thinks these machines are magic and watches them work like a TV show.

For as long as FDM/FFF 3D printers have existed, we've been trying to find ways to make our prints colorful. You can read about the existing landscape of multicolor systems on page 28, but once you get a printer, there's still a lot to consider.

While machines like the Bambu Labs printers with Automatic Material System (AMS) are pretty amazing, they waste a lot of material and time. Consider for a moment that every color change takes about a minute and spits out a "poop," and this can happen several times per layer. Because of these issues, it may still be to your benefit to approach multicolor more like we used to in the old days, by basically ripping our models apart and printing the parts separately.

Let's look at one example to break down the pros and cons of using the Bambu AMS versus breaking apart your model. I printed this model in one go on a multicolor Bambu A1 Mini with v2 of the AMS. I also printed it using various single-color printers to achieve the pieced-together approach (Figure Ⓐ).

AMS METHOD

PROS:
- Just hit print an d walk away.
- Control where you want certain colors — you can determine them at the time of slicing.
- Tiny pieces can be easier, as you're not printing a tiny thing on your bed (like the fog lights).

CONS:
- It's slow. This print took almost 4 days. It makes experimenting and tweaking settings very frustrating and wasteful — as you'll note, my support settings need adjustment but I didn't know that until after 4 days of printing.
- Material waste. The poops (Figure Ⓑ) and purge tower for this model were considerable.
- Lack of flexibility for orientation. Some parts could be crisper if printed separately at their own optimal orientations.

OLD METHOD

PROS:
- It's fast. Cumulatively these parts took roughly a day of print time. And I used four printers so I got it all done in only a few hours (that's

another pro, if you have multiple printers).
- Can choose optimal orientation for each item.

CONS:
- Gluing the parts together can result in less than polished presentation (Figure Ⓒ).
- Tiny parts — let's face it, even the nicest FDM printer doesn't really like printing itty bitty objects. For example, on this model the fog lights are maybe 3mm by 2mm.

There is no clear winner here; both ways have their strong points. However, in many cases you simply won't have the option of breaking your model apart unless you have the skills to do so. Some models are already broken apart, like this one, but as we move into the future of machines that can use systems like the AMS, we'll see that less frequently.

I think it's important to point out that a multi-toolhead system like the Prusa XL (reviewed on page 124) is really the best solution to this problem, resulting in less waste, more material options, and faster print time. However, those systems are still very expensive and quite rare. ✹

Assembling parts the old way can result in less than perfect fitment.

The AMS makes everything fit perfectly, but you better hope your settings are dialed in, because if you need to adjust things, you've got a long process in front of you.

FILAMENT PAINTING

EXTRUDE FULL-COLOR IMAGES EVEN WITH AN ORDINARY 3D PRINTER, USING HUEFORGE'S CLEVER SOFTWARE

Written by Caleb Kraft

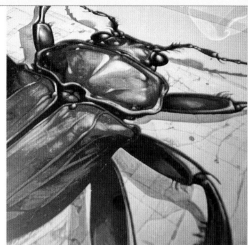

3D printing is always surprising us with new and inventive ways of using the somewhat simple machines. In this case, it's new software called HueForge (shop.thehueforge.com) that creates beautifully rendered 2D artwork using the very limited colors and materials allowed by 3D printing. You don't even need to have a multicolor setup to do it!

To understand what HueForge is and why it's so neat, we first have to break down the limitations of 3D printers when it comes to making 2D art:

- Filaments come in a limited range of colors, and you're typically limited to only four of them even when using a fancy auto color changer
- The nozzle is relatively large, usually 0.4 to 0.6mm which is huge compared to a pixel on your screen or on a page.

HueForge overcomes these two big difficulties by utilizing the semi-transparent nature of very thin plastic to achieve color blending effects. Using only a few colors, this software will put down layers — darkest beneath, brightest/most translucent on top — building up various areas to get gradients and transitions that look stunning. If your printer isn't equipped to do multicolor, you can still use HueForge to make incredible multicolor art — the machine will pause between each color, allowing you to manually swap filaments.

Frankly, it's rare to see a piece of software that really expands on the capabilities of a 3D printer and I think HueForge has pulled that off. ◐

Example 3D-printed artwork showing how the layers are built up, starting from right to left.

Ian Smalley, HueForge

The author shows a threaded bolt printed in about 10 seconds at Maker Faire Bay Area 2023.

3D PRINTING IN SECONDS

Written by Taylor Waddell

LAYERLESS, ALL-AT-ONCE RESIN PRINTING IS A REALITY WITH THE ASTONISHING CAL SYSTEM

TAYLOR WADDELL is a third-year Ph.D. student at UC Berkeley and a NASA engineer. To Taylor, makerspaces are home.

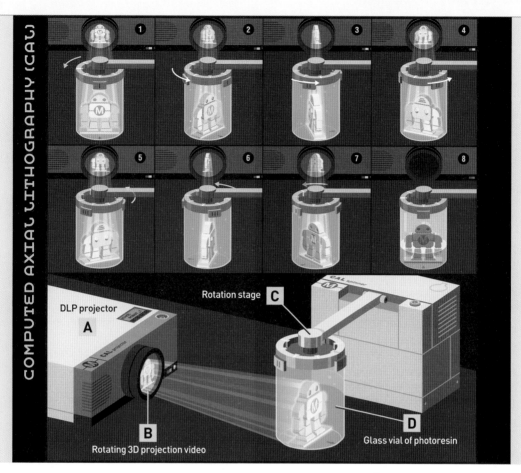

DLP projector **A**

Rotation stage **C**

B Rotating 3D projection video

D Glass vial of photoresin

"Any sufficiently advanced technology is indistinguishable from magic," wrote Arthur C. Clarke. And with *computed axial lithography*, or *CAL*, we've finally created magic.

CAL is a layerless 3D printing process invented at the University of California, Berkeley in 2019. There's no growth, no buildup — parts are made entirely at once. The common question, after a few moments of trying to comprehend what was just said, is "How?"

HOW IT WORKS

First is the printing material, typically a UV light-cured resin whose chemistry is similar to those used in stereolithography (SLA) printing. The simplest material is comprised of a *monomer* and a *photoinitiator*. A monomer is a molecule that can react to grow long molecule chains and intertwine with itself, creating a polymer. The reaction is started by the photoinitiator, a special molecule that will react to a certain wavelength of

light. The materials can get much more complex than this, having multiple monomers, or more molecules that react to light differently, or even tiny silica/glass particles. CAL has already been demonstrated with over 60 different materials.

Next is the mechanical setup of the printer. The illustration above shows almost exactly what the patent looks like. The printing material is put into a **clear cylindrical glass container D**. There's a **rotation element C** that rotates the container, and a **projector A** that shines the **images B** that will create the part. Everything else is extra and is just used to improve the printing process. This could include lenses that will better control the quality of the light rays, or an index matching box that will prevent the light from refracting when hits the edge of the rounded glass vial.

The magic of CAL is almost all within the images the projector shows, which have their roots in *computed tomography*, aka a CT scan. In a medical CT scan, a 2D fan of X-rays is directed

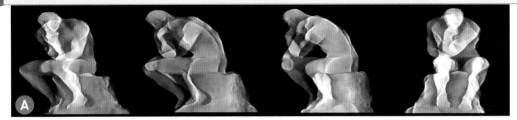

Light intensity to object density: A sequence from a rotating CAL projection video shows its ghostly quality — not a hologram, not a 3D model, but something new.

through a slice of a patient's body. Then the X-ray source is rotated to a small degree, and the process is repeated. After several rotations and complex math, a clean slice of this person can be **reconstructed** — computed by determining how the X-rays absorbed energy through the patient and then using some filtering techniques. Now the spinning X-ray source is moved up and down the person, to get multiple slices. Once the length of the person is fully scanned, the slices can be put together to make a full 3D model of the patient, insides and all.

CAL is essentially a CT scan in reverse. It starts with a 3D model and computes the slices, from multiple angles, that it would take to form that 3D model. This process begins with breaking down the model into **voxels**, little cubes that are essentially 3D pixels. Then, like a CT scan, a slice is taken from the model. Starting from 0 degrees, a one-dimensional row of varying intensity pixels is calculated, based mainly on how dense the object is viewed from that angle. This is repeated after rotating in small increments, for 180 degrees. (Not 360° — the row of pixels would be identically calculated when viewed from 180°, so there is no need to calculate twice.) The combination of these 1D rows of pixels from various angles is called a **sinogram**. Then, like in a CT scan, this is done for multiple slices of the model until sinograms are created for all slices.

Now the first version of the projected image can be made. By taking all the 1D rows of pixels from a certain angle (such as 30°) and stacking them on each other, a 2D image is created that accounts for 3D space. However, the first version of this image is going to be blurry and distorted in the wrong ways. Through filtering, optimization methods, and other math tricks, a high-quality 2D image is created. This image can then be projected into our vial, to start forming our part.

But to form the desired geometry, many angles of these 2D projections are needed, which are combined into a 3D **projection video** (Figure A). A typical video created for projection will have 180 unique frames/angles. This is just the start of how these videos are created. CAL also computes a lot of physics that is involved with the printing process — how the light refracts when hitting the surface, how molecules move around in the printing material when parts form, etc. — and the video can be modified and distorted to account for these effects. There are few physical effects that can't be corrected for.

PRINTS IN SECONDS

The three core parts of CAL are now complete: the material, the mechanical setup, and the special projection videos. Now begins the printing. First the vial will start rotating about its central axis, then the projection video is illuminated through the vial, with the focus traditionally being at the center of the vial. However, because light is passing through the entire volume of the vial at each angle, everywhere within the part is starting to be formed. The material needs a certain dosage of light before it starts to become solid. As the vial is rotated, only the spots where enough combined light has passed through will become solid. This truly causes the part to form all at once, sometimes in as little as 10 seconds.

CAL parts do have to undergo post-processing. The part is removed from the vial with tweezers, washed in a solvent to remove its excess resin, and then exposed to more light under a strong UV LED lamp to reach its full mechanical properties.

Because of the unique way the parts are formed, CAL can do several unique things, such as **overprinting**. *Overmolding* is when an existing part, such as a screwdriver bit, has plastic molded around it, creating a screwdriver handle. CAL can do the same thing with printing, forming 3D geometry over existing parts. The examples in Figure **B** were printed in just 20 seconds (rather than 1 minute), sacrificing quality for speed.

CAL also does not need support structures to create parts. As the part forms, it is upheld by the printing material itself; this is why CAL typically uses much more viscous materials than SLA. A common CAL material has viscosity similar to molasses, or even higher.

CAL IN ZERO GRAVITY

High-viscosity materials tend to have stronger and more desirable printing properties. However, if there's a need for lower-viscosity material, several different techniques can be used, the most exciting of which is simply removing gravity, by using CAL in space! CAL has already been demonstrated in microgravity (the more scientific term for "zero gravity"), on parabolic airplane flights (Figure **C**) where it printed over 400 parts. While the microgravity environments only last for 20 seconds on the parabolic plane, CAL is able to completely form parts. The next step for the space version of CAL is to be tested in suborbital space, and then eventually on a space station!

WHAT'S NEXT

Since the technology's invention in 2019 by Brett Kelley and others under the supervision of

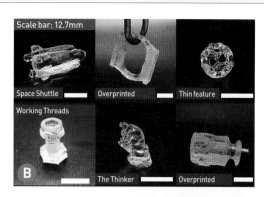

Scale bar: 12.7mm

Space Shuttle | Overprinted | Thin feature

Working Threads

The Thinker | Overprinted

ne green, "power" LED
ndicate th... trument i
isplay shows the curren
tart time date. ...00 sec
me the pump (and oth
...rn on, to begin the

Dr. Hayden Taylor, much has happened. Not only has the technology been tested in microgravity, but pure glass parts with feature sizes 5x smaller than the width of a human hair have been made! Much more is in the works as well, from optical lens printing (Figure **D**), to printing in a half-meter diameter vial, to recyclable materials.

CAL has grown beyond UC Berkeley as well, with more than a dozen institutions researching and expanding the technology. And it's an open source project, so you can try it yourself at github. com/computed-axial-lithography. The future of CAL seems limitless. ●

Written by
Joan Horvath

OPPORTUNITY
ACTIVATED

Diana Hall, president and CEO of ActivArmor.

ONE MEDICAL MAKER'S JOURNEY FROM POSSIBILITY TO PRODUCTION

JOAN HORVATH is the co-founder of Nonscriptum LLC, an MIT alum and recovering rocket scientist, and author of many books including *Make: Geometry*, *Make: Calculus*, and *Make: Trigonometry*.

Makers often look at a problem and think, *I can fix that.* But not many follow through to the point of creating a business, especially in a highly regulated industry. Diana Hall pulled off that sequence in about a year after recognizing a problem potentially solvable with a 3D printer.

It started when Hall was volunteering for an organization that supported children who were victims of abuse. Some of these children had broken bones, and often were in temporary living situations where hygiene was a problem. Their plaster casts got wet and dirty, leading to other health issues.

She wondered if 3D-printed casts would be a possibility. Trained as a chemical engineer, with an MBA and professional software experience, she knew just jumping into manufacturing was unlikely to be a feasible (or legal) option. After some discussions with a doctor also working with the children, she called the FDA to determine the regulatory requirements.

MAKING A MEDICAL DEVICE

Hall learned that a 3D-printed orthopedic cast would be considered a medical device, unlike plaster casts which were considered simply a process that a person could be licensed to perform. This was a significant change from the existing regulatory environment. After about a year of working through this thicket, this result was the ActivArmor cast.

Although the devices look simple, there are many challenges. "You have to consider biocompatibility," Hall says. "That means it won't be reactive to the skin. And there are ISO certifications, for example ISO 10993-5, which is the cytotoxicity test. They have to be done for medical devices both before and after manufacturing of the device."

But for 3D printing, it's challenging to do that, particularly if every single design is custom, even varying in how strong they need to be. That meant that each device had to be made by an approved process and facilities, using tests and standards worked out with regulatory bodies.

To create each custom cast, first a scan is made of the patient's injured body part, using ActivArmor's phone app. Then the device is

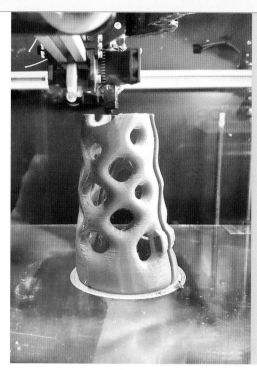

A 3D-printed cast can outperform traditional plaster.

created by an ActivArmor provider using the company's proprietary software, which has been a crucial invention. The cast can be locked on, or not, depending on whether the patient is allowed to remove it.

The material has to be not just biocompatible but also non-porous, so that customers can bathe, swim, and live fairly normally, versus a traditional cast that can't get wet. To accomplish that, the casts need to be post-processed to remove layer lines. If you're a 3D printing person, you know that the conditions in the room and even the batch of filament can affect the outcome, so ActivArmor and its providers need to manage those variables. And there are more issues in the case of a cast, such as transparency to X-rays.

MANAGING SUCCESS

As ActivArmor approaches its 10th anniversary, managing growth and maintaining regulated quality keeps Hall busy. She has also had to overcome an impression that 3D printing is a process that remains hobbyist-level. Early adopters may have tried medical 3D-printed products that lacked the investment and evolution Hall's enterprise has put into theirs.

ActivArmor

Casts produced at ActivArmor's manufacturing facility.

And naturally, medical technicians who make casts now are nervous about the change. Hall says, "What I'm telling them is that this is going to make your job easier, faster, more efficient, and your quality is going to go up. And your patients are going to be happier!" Printed casts will evolve these jobs, she says, but not eliminate them.

As a solo female founder, Hall has had her share of dismissive potential collaborators. When going into a meeting with advisors, she says, everyone would automatically focus their attention on the tallest male walking in with her. But she enjoys watching them shift to her when they realize she knows what she's talking about.

"The more we as women prove that it's not about what we look like," Hall says, the less frequently those assumptions will be made. She hopes that instead people will enter a meeting thinking, "OK, let's see who this person is."

Why has Hall stuck with the turmoil of starting such a challenging business, instead of a stable engineering career? "You make sacrifices in your life. But I also get to get up every morning and help people, and make a difference, and do something that I feel inspired to do." Some days she'll be ready to throw it all in, she says, but then, "One mom will send me a picture of a little girl on the beach and a note that says, 'Hey, look, you saved our summer.' And then I'll be like, OK, fine. And back to work!" ●

Unlike plaster casts, ActivArmor casts are waterproof.

Relaxing in the pool — summer saved.

ActivArmor

MAKEY MARK!

Written and photographed by Sujay Saravanan

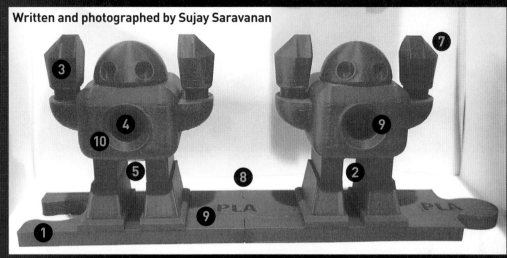

TRY A NEW BENCHMARK PRINT WITH A FAMILIAR FACE!

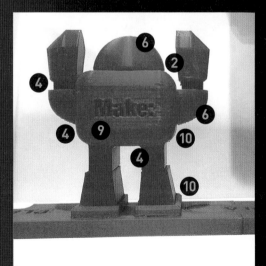

SUJAY SARAVANAN is a senior at Irvington High School and Berbawy Makers in Fremont, California, who loves all things 3D printing and design.

Makers, meet Makey Mark!

I've turned *Make:*'s mascot, Makey the robot, into a 3D-printed benchmark object for testing various qualities of 3D printers and filament types. Inspired by features on other benchmark objects, Makey Mark adopts some of those characteristics while keeping the classic robot's look. For example, the arms have a surface that ends in an edge, creating a shape with a round bottom and a triangular top. This mimics the bow of the classic 3D Benchy boat benchmark object, incorporating the iconic surface quality test into Makey's design.

Here are some of the tests that Makey Mark is good for:

1. Warping from the flat base
2. Stringing between vertical surfaces
3. Layer adhesion (durability of arms)
4. Overhangs
5. Bridging
6. Round and spherical surfaces
7. Sloped surface quality (like Benchy's bow)
8. Tolerance between puzzle pieces
9. Text engraving on both horizontal and vertical planes
10. Chamfers and fillets

Download Makey Mark from printables.com/model/715584 and let me know how you like it! ◢

SLICES
ARE READY!

TODAY'S MENU OF PIPING-HOT
3D PRINT SLICER SOFTWARE

Written and photographed by Courtney Blum

Carnival Flask by Clockspring3D in Bambu Studio.

COURTNEY BLUM aka Filament Stories is a short-form content creator talking with great enthusiasm about all the myriad colors and styles of 3D printing filament and keeping us in touch with the latest trends and sheer joy that a spool of plastic can be when turned into beautiful 3D prints.

3D modeling is used in a multitude of industries, but if you want to take one of those models and print it, you'll need a bit of software called a *slicer*. In essence, slicers are pretty simple. They take a three-dimensional model and "slice" it up into layers. What happens next in the slicer depends on which type of 3D printer the model is being sliced for:

- For filament-based printers, aka FDM (fused deposition modeling) or FFF (fused filament fabrication), the slicer's job is to further divide each layer into a **series of motions** telling the printer how to move and when to extrude, or print the plastic. Each of those steps is a line of *G-code* which, combined with general settings, make the roadmap the printer needs to print the model.

- For resin-based printers using stereolithography (SLA) or digital light projection (DLP) technology, the slicer creates a **series of images**, or *masks,* one corresponding to each layer of the model. In the printer each mask is projected in turn, exposing each layer to a curing light to solidify the polymer. These image masks, along with basic printing parameters, provide everything the printer needs to know to turn the digital into the tangible.

Today's slicers are anything but simple. They provide extensive customization to define how a model should be printed, sometimes with hundreds of settings — which sounds daunting, to say the least. The good news is that most slicers have predefined profiles for the most common settings, enabling users to get started knowing little more than how to load a model and select "Slice."

Here are some of the most popular 3D printer slicers currently.

FILAMENT-BASED SLICERS (FDM/FFF)

❶ ULTIMAKER CURA

Initially created for UltiMaker's desktop printers, Cura is open source, and so it's used by a large array of 3D printer brands who've customized and rebranded it for their printers. Cura and its variants are used by more people than any

Carnival Flask and Myriad Vessell by Clockspring3D

other slicer, supported by a passionate user and developer community.

- Cost: Free (Paid professional versions)
- Supports STL, 3MF, and OBJ file formats
- Print status and monitoring in-app configurable based on printer and setup
- Marketplace with community-based plugins to extend base feature set
- Slower slicing times than other products
- Frequent product updates

❷ PRUSASLICER

Developed by Prusa Research, originally from the Slic3r tool, PrusaSlicer currently has profiles for many brands and printers. Its intuitive interface creates a seamless experience from beginner to experienced user. As an open source product, PrusaSlicer has been used as the basis for other popular slicers.

- Cost: Free
- Supports STL, STEP, 3MF, OBJ, and AMF file formats
- Intuitive interface
- Fast slicing times
- No in-app monitoring
- Frequent product updates

❸ BAMBU STUDIO

A recent entry in both 3D printer and slicer world is Bambu Lab, who fast built an impressive collection of printers capable of multicolor printing (see pages 28 and 125). Bambu Lab took PrusaSlicer and made it their own, adding new features and improving on others, and recently integrated their model repository, Maker World, allowing you to select and print models directly from within their slicer.

- Cost: Free
- Supports 3MF, STL, STEP, SVG, OBJ, AMF file formats
- Project-based workflow
- Only supports Bambu Lab printers and some select brands
- Fast slicing times
- Intuitive model painting up to 16 colors
- Frequent product updates
- In-app status, control, and video monitoring of Bambu Lab printers
- Integration with Maker World model repository for direct printing
- Multiple print plates can be sent to different printers in a single project

❹ ORCASLICER

Considered by many to be the most advanced slicer available, OrcaSlicer has come on the scene recently. The power of open source is nowhere more apparent than here: Bambu Studio user SoftFever created Orcaslicer as a fork of the Bambu Studio, creating one of the most feature-rich slicers, adding new capabilities and settings and also rolling in functionality from SuperSlicer, which had stagnated in development.

- Cost: Free
- Supports 3MF, STL, STEP, SVG, OBJ, AMF file formats
- Extends functionality of Bambu Studio and features from SuperSlicer
- Many printer brands and printers supported
- Fast slicing times
- Growing in popularity quickly
- Not associated with a company that sells 3D printers

There are many other slicers for FDM printers. **Creality Print** offers a sleek interface and print

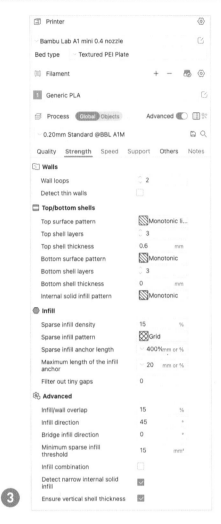

monitoring for Creality machines. **Kiri:Moto** is a browser-based slicer that provides platform independence, although its feature set is more limited. **ideaMaker** has cloud storage for files and also has interesting texture features that can be applied to models.

And we can't forget **Simplify3D**, a beloved slicer that was years ahead of its time. A new version was released in 2022, but it's hard pressed to stand up against all the free, richly featured slicers available today.

RESIN-BASED SLICERS (SLA OR DLP)

Slicing a model for printing in resin has several notable differences. Models aren't supported with a percentage of infill but instead are

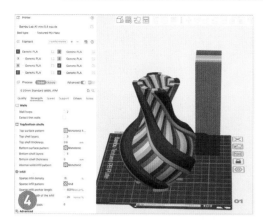

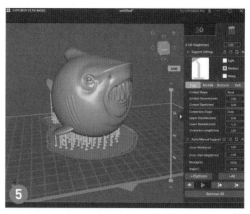

hollowed out, with drain holes added at key locations so uncured resin won't remain trapped inside. Because resin prints are printed upside down, support structures are key to success. The algorithms in resin slicers today are good at providing these supports — something very helpful to those new to resin printing.

⑤ CHITUBOX

Chitubox is the standout leader and the most popular resin slicer. The interface is user-friendly, with lots of settings that can be easily adjusted. Generation of supports has many options, including fully manual placement of supports. The free version is all most users need. Chitu produces motherboards for a number of resin printer brands.

- Cost: Free (Pro $149 per year)
- Supports a large range of printers
- Model hollowing well implemented
- Supports different resins with profiles for speed, exposure time, etc.

⑥ LYCHEE SLICER

Mango 3D's Lychee Slicer has been a favorite of many since its release. It goes beyond the functionality seen in Chitubox. The optimal part orientation feature helps reduce the amount of supports needed. Has added FDM slicing recently.

- Cost: Free (Pro $6/month, Premium $10/ month)
- Optimal part orientation
- Automatic supports well implemented
- Some of the best features are in the paid versions

⑦ PRUSASLICER

PrusaSlicer's easy-to-use, intuitive interface brings automatic support generation and part hollowing. But support is for Prusa resin printers only, so sliced files need to be exported as STLs and then brought into another slicer compatible with the user's printer.

- Cost: Free
- Second step needed to slice model after supporting/hollowing in PrusaSlicer for non-Prusa printers

⑧ PIKASLICE

For those who want to slice models from their iPhone or iPad, Pikaslice is a full-featured app that provides everything you'd expect. It's free to use, but it shows ads prior to slicing the model. There are various paid versions for both resin and filament slicing.

- Cost: Free ad-based (Paid plans for resin and filament)
- Fast slicing times

So which slicer is the best? What would I recommend? The better question is, what don't I use? They're all quality products. Some are designed to work with specific printers, some have features not found anywhere else. But the vast majority of features I expect in a slicer are available across the board, something we have open source to thank for.

If you're new to 3D printing, don't be afraid to try out a new slicer after you get comfortable in your first one. You might find you like another one more. ◗

CAN AI MAKE 3D PRINTS?

KIND OF. TEXT-TO-3D-MODEL GENERATORS ARE HERE, BUT NOT OPTIMIZED FOR PRINTING

Written and photographed by Caleb Kraft

Text-prompted 3D model from Luma AI, image fully colored.

A

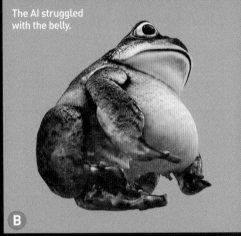

The AI struggled with the belly.

B

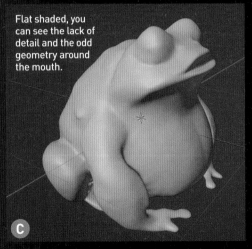

Flat shaded, you can see the lack of detail and the odd geometry around the mouth.

C

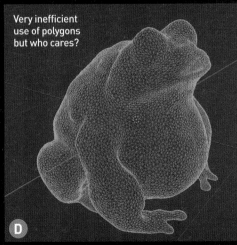

Very inefficient use of polygons but who cares?

D

CALEB KRAFT has been 3D modeling since the late 1900s and is still only barely better than this rudimentary AI.

With all the hubbub around generative AI, it isn't a stretch to start wondering in what new areas of making we might see this stuff proliferate. You can easily have ChatGPT write text for you or analyze your writing. You can instruct Midjourney, Dall-E, and other image generators to draw highly detailed, pixel-perfect creations in a variety of styles. What about 3D printing though? Can you type into a text box and obtain the perfect custom 3D printable model? Right now the answer is: kind of. However, in the very near future, that answer might be a resounding yes.

As of winter 2023–24, there really aren't any systems advertised with the intent of 3D printing, so I'll talk about the general concept of *text to 3D model.* This goal was out of reach a year ago when we published our guide to "Generative AI for Makers" (*Make:* Volume 84, makezine.com/go/makers-ai). In the short time since, the landscape of AI has been changing extremely fast and now we have a few different options for playing with text-prompted 3D model generators.

Ultimately, these tools are primarily focused on video game assets, so there are issues with 3D printing. While they do technically work, what you'll see is that the current generation of AI model generators relies on the color layer to convey many details that simply will not exist when you 3D print. This means your print may be blobby, lacking details, or even oddly formed.

There are a few places where you can try this kind of thing, such as 3DFY.ai, Sloyd (sloyd.ai), Masterpiece X (masterpiecex.com), and Luma AI (lumalabs.ai). Since Luma is free and easy, I tried it.

TEXT PROMPT TO 3D MODEL

In Figure **A** you can see the results of the prompt "cute toad, pixar style, studio ghibli, fat." (Don't judge me, I know what I like.) The textured version looks OK from certain angles, but we can see the feet and belly have some issues (Figure **B**), and fine detail is lacking (Figure **C**).

3D MODEL TO 3D PRINT

I had to convert the GLB file that was output by Luma AI to an STL file using Blender (Figure **D**), but aside from that, it was ready to print. What you see in Figure **E** is the result of a successful

E The toad model 3D printed. Technically doable, but much room for improvement.

print from my Bambu X1 Carbon.

While we can now say that we have used AI to generate a 3D printable model, we can also see that the geometry around the belly is very messed up. Printing it this way resulted in trapped supports that caused a mess when trying to remove them. I could bring this into modeling software and rebuild the feet and belly but at that point, with those skills, what do I need the AI for in the first place?

We've already seen 2D AI generative tools built into laser cutter software such as the xTool Creative Space. As these 3D tools improve, I can envision a near future where this kind of AI is built into slicers. Very soon you might just open your slicer, tell it what object you want, pick the best result, and hit Print! ●

PULTRUDER ALERT

Written by Joshua R. Taylor

MAKE YOUR OWN FILAMENT FROM RECYCLED SODA BOTTLES — USING SALVAGED PRINTER PARTS

JOSHUA R. TAYLOR dreams of making the world a better place with 3D printing. In 2019 while working with M3D's Crane QuadFusion, he started JRT3D.com to help the community troubleshoot and sell small parts. Frustrated with the waste that multi-material printing was creating, he switched to 3D printing recycling efforts.

PET is a fantastic plastic for 3D printing— it's tough, dimensionally and thermally stable, food-approved, and has a glass-like transparency (see *Make:* Volume 85, "Transparent 3D Prints"). PET is also one of the main waste plastics littering the planet, in the form of beverage bottles. Luckily there's a creative community of 3D printing fans who are dedicated to recycling waste PET into new printer filament.

PET pultrusion is our method of measuring a plastic PET #1 bottle's thickness, stripping it down to the appropriate width, and then pulling that strip through a heated 1.75mm diameter nozzle, reforming it into usable raw PET1 printer filament (Figure Ⓐ). It's the easiest, most affordable way to make DIY 3D filament at home!

Filament is commonly made by extrusion — a machine augers fully molten plastic to squeeze it out of a nozzle. PET pultrusion heats the plastic only to 210°C, reforming rather than fully melting it. This keeps it relatively safe and easier to use. We've created a pultrusion machine to do this: The ReCreator 3D (recreator3d.com).

The average 2L soda bottle is 0.30mm thick. When cut into an 8mm-wide strip and then pulled through the 1.75mm nozzle, it will reform to that diameter (Figure Ⓑ), yielding around 25 grams of filament, or about 1 hour of print time.

LOW-COST DIY FILAMENT

Using our plastic waste opens up low-cost options for DIY filament. Places like India and Africa struggle with their plastic waste, and they also struggle to import filament and printers in a cost-effective way. Our $20 filament in the USA costs around $60 in Africa, and our $99 Ender 3 specials at Micro Center would be $800.

If we're going to reduce our carbon footprint, it's crucial to change how we treat our waste. Recycling programs differ vastly, if they even exist. Some U.S. states pay 5 cents a bottle, others 10 cents, and some pay nothing at all. Germany on the other hand, a world leader in recycling, pays the equivalent of 25 cents on all plastic bottles, not just carbonated drinks!

BENEFITS OF PET

With PET pultrusion, not only do you get to recycle in a new creative way, you get to experience the

3D printer filament created from recycled PET soda bottles by the pultrusion method.

Close-up of PET bottle plastic, reformed into a straw-like strand of hollow filament.

PET1 mounting bracket outperforms PETG.

easiest way to make filament, and with one of the strongest, heat- and weather-resistant materials around! Some users have even created hybrid materials of PET1/PP and PET1/HDPE that can't be bought on the commercial market.

One benefit of using raw PET1 vs. PETG is its heat resistance. For the ReCreator 3D's heat block mounts, initially we used PETG but over time we saw the pultrusion process would melt the metal bracket forward. Switching to raw PET1 made them more heat resistant and less prone to melting (Figure Ⓒ).

And it's strong. We sent Stefan at CNC Kitchen some filament created on a PetBot pultruder.

Joshua R. Taylor, Stefan Hermann

The ReCreator 3D

D The ReCreator 3D MK5 Kit for converting an old 3D printer into a PET filament pultruder.

Stefan did a wonderful write-up (cnckitchen. com/blog/how-strong-is-pet-bottle-filament), performed tension tests, and covered it all in a video about the strengths of PET1 filament.

THE RECREATOR 3D

The ReCreator 3D pultrusion machine (Figure **D**) was created not just to re-create PET1 bottles into usable filament, but to do it by converting inexpensive and discarded 3D printers before they become e-waste. Already 3D printers are becoming items that are thrown away as freely as plastic bottles! Not only does this unit help stop plastic waste, it also helps us look at our aging technology for new functions.

Inspired by the PET-PULL group from Russia who first used the pultrusion method, I built the original ReCreator 3D based on an Xvico X3S 3D printer, using only its base structure, a few accessories, and a handful of 3D-printed parts. Since then the machine has had five revisions.

The ReCreator 3D MK5 Kit is made to be a universal solution that can make use of most standard aluminum extrusions in I3-style FDM printers, particularly the most affordable printer, the Ender 3. Many new users become discouraged and send these printers slightly used back to Amazon, which finds it easier to sell them at a discount than to repair and resell them for very little profit. The printer becomes an inexpensive throwaway or is resold for a very low price, if not given away as a gift.

This makes the ReCreator 3D MK5 Kit Ender 3 one of the cheapest options available for PET1 pultrusion, and the easiest way to reclaim waste as usable filament. For under $220, these units can be created by average people and turn homes into small recycling centers — as well as manufacturers using 3D printing.

FILAMENT JOINING

Filament segments can be joined by various methods that range from free upward to $500.

Start with a low-cost method: Use a candle to get the PET1 into a molten state and then pressure-force the two parts together inside a PTFE tube. This method can be counterproductive — if a seam becomes crystallized due to slow cooling, the join can easily break and cause extrusion issues inside the nozzle.

If you're considering using PET1 commercially and printing beyond 25g prints, the easiest way is to use a filament runout sensor. There are also options like the Enraged Rabbit Carrot Feeder (ERCF), which uses six different filaments for continuous printing. The **Mosaic Palette 2 Pro** is one of the most reliable joining options (see page 30). The ReCreator 3D spent 6 months with Jonny at Mosaic Manufacturing, testing PET1 filament. They found that the Palette 2 Pro was the best solution from their product line. And I recently reviewed the **Slunaz FF1R Kit** (Figure **E**), an

E

F

Filament from soda bottles suggested slicer settings using CURA 4.8 and Ender 3

- Layer Height..............0.2mm
- Nozzle Temp.............260
- Bed Temp.................75
- Flow..........................130%
- Initial Layer Flow.......130%
- Print Speed 30mm
- Fan............................15%
- Retraction.................4.5mm
- Retraction Speed.......40mm

excellent Kickstarter machine coming in at half the cost and created specifically for joining PET1 and other small-yield leftover filaments (jrt3d.com/slunaz).

PRINTING WITH PET1

Print quality is consistent and reliable. We are creating our filament from waste, so it may not be perfect (air bubbles, slight imperfections). We're going to have 95% quality, but this can be the case with even commercially produced filaments. Due to its high-temperature nature it does need a printer that can reach 260°C (Figure **F**). This however allows us access to a very robust material. It's great for rapid prototyping and for projects that need to resist heat or weathering.

GOING FURTHER WITH PET

PET pultrusion isn't without its own waste. Tops and bottoms of bottles are typically discarded. One of my goals is to process this waste into viable material. I'm now using a **mini shredder** (Figures **G** and **H**) from DIY Chen (diychen.net) and a **filament extruder** (Figure **I**) from Artme 3D (artme-3d.shop) to make 3D filament from my waste material.

I also aim to create a DIY **direct drive extruder for shredded particles** (Figure **J**), working with other leaders in the pultrusion community. Join us at **PET Pultruders United** (ppu3d.com) as we continue to grow and learn from one another. ⊘

G

H

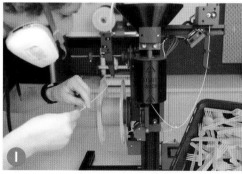

I

J

Stefano Lunazzi, Joshua R. Taylor, DIY Chen, Stefan Hermann

DIODE POWER!

NEW DIODE LASERS RIVAL CO$_2$ CUTTERS FOR POWER — AND LAST FAR LONGER

Written by Caleb Kraft

Multilayer Cat in xTool Creative Space by user cc45.

Caleb Kraft

I used to say diode lasers were "throwaway lasers." They were so underpowered that I felt that they were almost pointless, especially when you could buy a K40 — a common, super cheap CO_2 laser cutter — for $350–$600. Back then we would regularly see diode lasers touting 6 watts of output power on flimsy frames. Compared to the 40 watts (realistically 20W–30W output) of a K40, they just didn't seem worth the time. Over the past year or two, this has changed.

Diode lasers can be stacked to increase the tool's power output, but still, for a long time that only pushed into the teens. Now we're seeing machines pushing 30 or even 40 watts of laser output — fully capable of competing with a K40 or even something like a Glowforge in terms of raw power. Sure, the diodes are a different wavelength of light than a CO_2 laser, and therefore have some differences in capability — diodes struggle with transparent materials — but the differences might not affect you as much as the benefits. A diode can last virtually forever whereas a CO_2 tube has a lifespan measured in thousands of uses. This difference in lifespan makes pricing feel a bit imbalanced, as the diodes are priced higher than a K40, but keep in mind that you're not going to have to replace that diode, clean or align mirrors, or deal with a cheap high-voltage power supply.

Two diode lasers have popped up recently in the same power range as the typical K40 — probably more by the time this gets published — and frankly, it's nice to see the K40 get some real competition! As these get more common, we suspect the price will drop accordingly.

❶ ACMER P2

The Acmer P2 boasts 33W of output. That's pretty impressive for a diode laser. This system does have an open gantry, but still manages to include built-in air assist for nice clean cuts. I would consider this a direct replacement for a K40. Selling at a little over $1,000, it is a bit pricier, but again you aren't going to have to replace or align mirrors or the tube.

❷ XTOOL S1

The S1 from xTool is touted as a 40W system. This one also includes a full enclosure, which is not only pleasant but required for some school environments. Enclosures typically bring safety features like auto shut-off, as well as allowing for the fumes to be extracted more easily.

This machine has features, such as robust design software, auto focus, and curved surface compensation, which combined with the rock-solid construction make it *feel* like a much nicer tool. The only thing that makes it feel less capable than a Glowforge or Muse is the lack of a camera, and at nearly half the cost of those other two, I'm willing to forgo that feature! Even with the cost of the optional IR diode head, which would greatly expand the capabilities of this machine, it is still nearly half the cost of the comparable CO_2 machines. ◗

TOP TIP:
IR LASERS FOR METAL

Last year we saw the introduction of infrared (IR) diode lasers that could mark metal directly, such as the xTool F1, which has a dual laser system. If you need to mark metal, you should look for a laser engraver that does IR, as these will be much cheaper than purchasing a stronger fiber laser.

DIY Graffiti Projector

Written and photographed by Claire Danielle Cassidy

CLAIRE DANIELLE CASSIDY is an open source electronic hardware project manager, full stack web developer, artist, designer, and small business owner. Her dream job is teaching a diverse range of people about technology that makes a positive impact.

Make a handheld graffiti gobo projector, and let them read the lighting on the wall

"Guerrilla projection, pioneered by artists and advertisers, has been increasingly embraced by activists in recent years as a new medium for delivering messages. The advantages are obvious: With a single high-powered projector, you can turn the side of a building into a huge advertisement for your cause, plastering your message on a spot that would otherwise be out of reach. It's legal, relatively cheap, and ... most importantly, it's visually powerful: You can literally shine a light on the opposition."
—from Beautiful Trouble, edited by Andrew Boyd and Dave Oswald Mitchell

TIME REQUIRED: 1–2 Hours
DIFFICULTY: Easy
COST: $45–$60

MATERIALS
Really all you need are these four things. Important details about each of them are available in their respective sections of the tutorial.
- » **LED flashlight, 1,000 lumens** We used the SupFire C8-T6 Tactical Flashlight, Amazon B07FBJBZ25.
- » **Lens from projection TV** Search eBay for "rear projection television lens."
- » **Stencil** of whatever words or image you want to project
- » **Coupler** to connect it all. We used a rubber plumbing coupling, 4" to 1½", Fernco P1056-415, Amazon B00011AVCS.

TOOLS
- » **Laser cutter (optional)** or other way to cut your stencil
- » **Screwdriver** to tighten the coupling clamps

Got something to write on the wall? Write it big — with light!

This tutorial is here to help you build a low-profile handheld light projector from relatively accessible hardware store parts. The project should cost around $45–$60 in materials, but might be cheaper (or pricier) if you find parts on your own.

There could be a thousand creative ways to do this type of projection. We have worked hard to give details for the specific build we know works, but we hope that people will also try to design similar, not exact copies. This is because the exact parts we used may be difficult to source in the future, though similar parts should always be available, and because alternate builds could have advantages of their own, such as lowering cost or improving projection throw.

This project is also completely open source, meaning that you can make, share, use, alter, and yes, even sell this design or any similar design, with no credit or attribution needed. You can (and we encourage you to) fork this project on Github. This information is yours.

HOW DOES IT WORK?

What you're doing in this project, essentially, is sending a flashlight's beam farther and more sharply, using a big lens, than it can do on its own. Then you're adding a message with the stencil. Putting them together with a coupling is all about finding the focus — the spacing between the flashlight and lens.

This type of projection is called *gobo projection*, and is often used in theater and stage settings. Gobo stands for "goes before optics" because putting a stencil between a light and a lens projects a crisp, focused image. A stencil used in this way is sometimes referred to as a gobo, and can be made of clear acetate sheet, acrylic, wood, or metal — depending on how hot your light source gets.

BUILD YOUR HANDHELD GRAFFITI PROJECTOR

Before you start, please watch our tutorial video at vimeo.com/851967606 for an overview of how all the parts should fit together.

PROJECTS: Handheld Gobo Projector

1. FIND YOUR FLASHLIGHT

Since we created this tutorial at github.com/
DisruptivelyUseful/handheld-light-graffiti,
flashlight technology and parts have evolved at
an extremely rapid pace. We try to keep linking
to flashlights that will work for this build, but it's
hard to keep up. This means it's important for
you to understand the *properties* of the type of
flashlight that will work for this project, so you
can try out ones you can find yourself.

Really any light source can work for this type of
projection, but LED flashlights are the cheapest,
smallest, safest option as they are very bright,
portable, and rechargeable, and don't produce a
lot of heat that could melt or ignite your stencil.

Brightness: In the build shown in our video
we used a 1,000-lumen "tactical" flashlight
we got on Amazon a few years ago for around
$20. The brighter the better, but a 1,000-lumen
flashlight with a fresh 18650 battery can throw a
remarkably large and clear image in reasonable
darkness. We have found essentially the same
flashlight from a few different suppliers since
then, and we suggest you look around yourself.

Batteries: Flashlights that use rechargeable
lithium 18650 type batteries are a great option,
because six or eight usable 18650 cells can be
recovered from old laptop batteries, and a single
18650 can power one of these flashlights for
4–7 hours. (It's OK if your flashlight uses AA or
AAA batteries too, we just like lithium.) You'll
probably want extra batteries with you to last the
night. Having a 18650 battery charger and extra
batteries is a great investment. These are easily
found online.

Reflector: It doesn't matter so much which brand
of flashlight you use, but we learned the hard
way that some LED flashlights use a **parabolic
reflector** to focus the light, and others use a
clear plastic dome-like lens. Both work fine for a
flashlight, but for use as a projector the ones with
parabolic reflectors work *way* better, because they
hold the stencil much further away from the hot
LED and because they don't complicate the optics.

The ones with (good!) parabolic reflectors look
like Figure Ⓐ. Notice the shiny, concave internal

surface with flat clear plastic cover. This is what
you want.

You don't want one that looks like Figure Ⓑ.
Notice the lack of shiny interior, and the convex,
bubble-like clear plastic dome cover. Not great
for projection.

Size: The head of the flashlight we used fits
perfectly into a 1½" plumbing fitting designed for
ABS pipe, which made coupling it to the lens easy.
Check the diameter of the head of your flashlight,
and make sure it will fit into the smaller end
of your plumbing coupler. The flashlight we
currently suggest is the SupFire C8-T6 Tactical
Flashlight, which comes with an extra 18650
battery and battery charger.

If your flashlight does not fit into your coupler,
try wrapping several layers of fabric, tape, or
other material around the flashlight until it fits
loosely enough to move, but tightly enough to stay
in place. We found that long, thin strips of velcro
worked very well for this.

2. MAKE YOUR STENCILS

You'll need to be able to make gobo stencils to throw your own images. We used a Glowforge basic 40W laser cutter to cut thin (¹⁄₁₆") black acrylic for our stencils, which worked well with the style of LED flashlight we suggest. We made a round stencil at 1.61" diameter to fit inside the head of the flashlight we used. We have provided these SVG files in the Github repo for you to use. They may (probably) need to be resized to fit your flashlight. If you don't have a laser cutter you could ask your local makerspace for access to one, or use an online laser cutting service like Ponoko.

Another, more accessible way of creating stencils is by printing transparencies, which we have provided PDF files for. Again, you might need to resize them to fit your flashlight.

You can also experiment with cutting your own stencils using materials like thin plastic, stiff card stock, or even a tin can and an X-Acto knife.

Cut-out stencils work best for larger words and simple phrases. For smaller/more text/detailed images, we found that laser engraving an image onto transparent acrylic, and then spray painting the engraved area black before peeling off the protective paper, worked surprisingly well. It may be worth playing around with that approach if you do have access to a laser cutter.

> **CAUTION:** Test your stencil beforehand and make sure it isn't getting too hot for the material you are using! This is particularly important if you are using paper products to make your stencil.

Some other approaches you could try:

- Make a low-tech DIY metal gobo, from a tin can or pie plate: instructables.com/id/Home-made-Gobos
- Draw on a sheet of transparent plastic with a Sharpie.
- Cut a stencil out of cardboard with a razor blade, and add tinfoil if it's getting too hot from the LED flashlight.
- 3D print a stencil. You can search Thingiverse and similar repositories for examples.

Get creative! Maybe let us know what works well for you!

eBay

3. SCROUNGE A LENS

Nearly any lens of any kind will work to some extent, but some will work (much) better than others. All lenses have a *focal length* — the distance from the lens where the image comes into focus. We used a lens from an old big screen projection TV, because we had a few lying around. They work well, and three of them can be found inside any old boxy projection-style CRT TV. If you take off the screen, it should look something like Figure C.

> **TIP:** If you're unsure whether a TV is the right kind, try gently poking the screen. Is it large, plastic, textured, and a little floppy in the middle? If so, you're in luck!

Not everyone wants to rip apart a big screen TV, however, so we did some research and found that you can buy this style of lens on eBay for $5–$20. Search for "rear projection television lens" and look for a lens that looks like Figure D. You can also ask around for old TV lenses (and camera lenses, see below), and check your local thrift shops and re-stores.

Each lens comes in two parts: the lens assembly, and a plastic housing that slides along two screws, allowing for fine adjustment of focus.

For our purposes, we won't need the outer housing. Just unscrew the two screw posts and pull the lens away from the housing (Figure **E**).

DANGER: Never leave a big screen TV lens out in the sun. It can act as a giant magnifying glass and light things on fire when the sun hits it!

The lens assembly has a stepped shape that makes it easy to find a step with the right diameter to couple to a plumbing fitting. We found that the second step on ours was almost exactly 4", and fit a 4" rubber fitting for ABS pipe.

Camera lenses: The now-defunct GuerillaBeam project (streettoolbox.fandom.com/wiki/Guerilla_Beam) used a 3D-printed slide holder to couple 35mm camera lenses to LED flashlights. Old 35mm lenses can usually be found at thrift stores for $5–$15. They have very high-quality optics, and many of the bigger telephoto-style ones will let you make very fine adjustments to the focus and size/throw of your image.

If you use a camera lens, you'll have to play around with the spacing between the light, stencil, and lens to find the sweet spot, and then design your coupler accordingly.

4. CHOOSE A COUPLER

The coupler can be anything that securely holds the flashlight to the lens, and allows you to make small adjustments to the focus. We used a 4"-to-1½" rubber plumbing fitting meant for ABS drain pipe, Fernco P1056-415, which happened to fit our lens on one end and our flashlight on the other (Figure **F**). These should be available at any large hardware store. If your flashlight diameter is larger, search for a 4"-to-2" coupler. We prefer the high quality of the Fernco couplers because they are flexible and durable. Cheaper hard plastic couplers are very hard to work with.

These kinds of fittings come in a wide range of sizes, and are a good, cheap option for connecting things together. The distance between the lens and the stencil is more important than the distance between the light source and the stencil. Find the focal length by simply holding your lens in front of your stencil/flashlight, and then source your coupler accordingly.

E

F

Disruptively Useful

Since the coupling is all about attaching the flashlight to the lens with the correct distance between them to focus, we think a number of other things could work for this, though we haven't personally tried these:

GuerrillaBeam coupling: The GuerillaBeam 3D-printed slide holder/coupling (Figure **G**), was meant for a 35mm camera lens but a number of adaptations have been made, and a lot of these can be found when you search "guerillabeam" on Thingiverse or similar 3D model repositories.

Crafty coupling: Try cardboard, tape, hot glue, foam, toilet paper

G

tubes, pool noodles, etc. to figure out a coupling for your flashlight and lens if you want to get creative with cheap supplies.

LIGHT IT UP!

That's it! You're ready to shine a light on things.

Unscrew the cap of your flashlight and add the stencil. It should appear to be spelled backwards.

Insert the flashlight into the coupler and tighten the collar clamp until the fit is snug but the flashlight can still slide in and out.

Insert the lens into the coupler and tighten the clamp securely.

Adjust focus by sliding the flashlight in and out, changing the distance between the light and lens.

When you're happy with your focus, tighten it all together, and take it to the streets! Just remember, with great projection power comes great responsibility. ◕

> "This technology is very powerful, 'spectacular' in nature, and often under the control of one person or a small group who could potentially manipulate a large and impressionable crowd. This power needs to be kept accountable to the broader group, and should be wielded with great care."
> —Beautiful Rising on Guerrilla Projection

OTHER NEAT GUERILLA PROJECTION RESOURCES

- Beautiful Rising on Guerilla Projection as a protest tactic: beautifultrouble.org/toolbox/tool/guerrilla-projection

- Illuminator Collective have been light projection activists since 2012 and make tutorials: theilluminator.org

- Overpass Light Brigade make light-up letter panels: facebook.com/OverpassLightBrigade

- If you have the funds, you can also buy a professional gobo projector; look for portable, battery powered models.

- You can follow us, the creators, if you'd like: linktr.ee/ClaireDanielleCassidy and www.samsmith.work

Build-a-Bot (Cuddly Edition)

Written and photographed by Debra Ansell

DEBRA ANSELL is a maker and educator who will never stop demonstrating that LEDs improve everything.

Code and sew a plush, interactive, wearable companion robot

Hardware builds are seldom described as "touchable," "cuddly," or "cute," but I've found that embedding programmable electronics in soft accessories like handbags ("Bag to the Future," *Make:* Volume 87) and pillows ("Pixelblaze Pillows," Volume 83) produces projects with surprising tactile and visual appeal. Inspired by some clever and creative robot builders, I've experimented with placing controllers and peripherals inside plush sewing patterns to create soft, interactive companion bots. The results have been irresistibly cute and undeniably cuddly.

Make: has interviewed several talented makers whose companion robot builds started me down this path, and whose palpable enthusiasm provided the incentive to follow in their footsteps. Creator Jorvon Moss (makezine.com/article/digital-fabrication/me-and-my-robot-odd-jayy-interview) has designed an entire collection of clever robots with engaging personalities. Video host and hardware maker Alex Glow built an owl named Archimedes that sits on her shoulder and uses AI to respond to facial expressions (makezine.com/projects/google-aiy-robot-companion). And maker/educator Angela Sheehan presented her adorable color-changing plush dragon, Nova, at the recent Hackaday SuperCon conference (makezine.com/article/maker-news/hackaday-supercon-2023-lights-up-pasadena). Nova's lifelike movements and appealing softness were direct inspirations for my projects.

This article's featured bot, Rosie the Radiant Rodent, is a good way to learn about making a mechanized soft companion. She packs a plethora of functionality into a deceptively small package. A 9g servomotor powers Rosie's lifelike head motions, and three short "pebble" LED strings display colorful animations in her ears and tail. The two capacitive touch sensors inside Rosie's paw and head detect even slight contact through her fabric shell, while a Useful Sensors "Person Sensor" makes simple work of detecting faces in her field of view.

Rosie's responsive personality results from CircuitPython code running on an Adafruit Feather RP2040 controller, allowing her to respond to input by wiggling her head and

TIME REQUIRED: A Weekend

DIFFICULTY: Intermediate

COST: $30–$90

MATERIALS

» **Flexible "pebble" LED strings, 5V, 15mm spacing** AliExpress 3256805296568805. You'll cut three 7-pixel lengths.
» **Machine screws and nuts, M2: 8mm (2) and 12mm (4)**
» **3D printer filament, PETG (best) or PLA**
» **Adafruit Feather RP2040 microcontroller**
» **Useful Sensors Person Sensor** SparkFun 21231
» **Capacitive touch sensors, TTP223 (2)** such as Amazon B07K72N79J
» **Silicone covered stranded wire, 30 AWG** in assorted colors, Amazon B01M70EDCW
» **Hookup wire, solid core, 22 AWG**
» **Micro servomotor, 9g** Adafruit 169
» **Fleece/minky fabric: saturated color, 28"×9", and white or cream, 17"×9"** or larger, for body and for ears/tail
» **Thin cotton quilting fabric, 6"×6"** or larger
» **Thread** to match body fabric
» **Nylon coil zipper, 5"** Amazon B09TLBCHZB
» **Gaffer's tape** or other strong fabric tape
» **Heat transfer iron-on vinyl: black, 2"×2", and white, 1"×1"** or larger. Or substitute non-fraying fabric and double-sided fusible web.
» **Polyester fiber fill, 1 bag**
» **Large nylon zip-tie, at least 5mm wide**
» **JST-PH cable, female connector** Adafruit 261
» **DC power adapter, female 2.1mm barrel jack to screw terminals** Adafruit 368
» **Cable, USB-A to male barrel** Adafruit 2697
» **Heat-shrink tubing** narrow (1–2mm) any color, and medium (8–10 mm) clear
» **Stemma QT cable** Adafruit 4210
» **USB battery pack** Adafruit 1959
» **Servo extension cable** Adafruit 972
» **Clear nail polish**

TOOLS

» **Sewing machine** standard and zipper foot
» **Scissors: standard, fabric, and embroidery**
» **Sewing needle, straight pins, sewing clips**
» **Clothes iron**
» **Ruler and hole punch, 5mm**
» **Small screwdrivers** slotted and Phillips
» **Soldering iron, solder, and wire cutters**
» **Computer and printer**
» **3D printer (optional)** or service like Ponoko
» **Vinyl cutter (optional)**
» **Sharpie or fabric marker**

changing LED animations. She even reacts to nearby admirers by lighting up whichever ear is on the same side as the closest detected face.

BUILD YOUR SOFT COMPANION ROBOT

1. READ THE RAT PLUSH PATTERN

The sewing tutorial and pattern for a rat plush animal are at cholyknight.com/2018/09/28/rat-plush. Download the PDF and read all instructions from beginning to end before proceeding.

The pattern instructions are clearly numbered, with lettered sub-sections. This tutorial does not reproduce the existing sewing directions but notes the steps where you should modify or deviate from the original instructions.

> **HINT:** If you're unfamiliar with sewing stretchy materials or navigating a sewing machine around tight curves, you should first sew the pattern as written to familiarize yourself with the required techniques.

2. CUT PATTERN AND ATTACH FACIAL FEATURES

Print and cut the pattern pieces from fleece or minky fabric as directed by the sewing pattern instructions. Use saturated color plush fabric for the body, and white or cream fabric for the tail and inner ear. Select the ear pattern piece without the bite mark.

In addition to the original pattern pieces, cut a second body bottom piece (pattern piece F) from thin quilting cotton (Figure 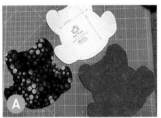). Any color of cotton is fine, as this additional piece will not be visible.

Pattern instructions **1** and **2** describe fusible web applique to attach fabric facial features. Depending upon the adhesive strength of your fusible web, these pieces may need reinforcement stitching around the perimeters. In my builds, I chose to make Rosie's features from glitter heat-transfer vinyl instead of fabric, importing the eye and nose outlines from the PDF pattern into my vinyl cutter software (Figure 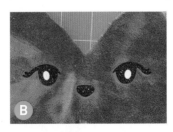). You can use either method; a vinyl cutter can cut both fabric and heat transfer vinyl precisely, but if you have a steady hand, you can hand trace and cut the features with sharp scissors before ironing them in place.

After the facial features have been ironed on and stitched if necessary, use the 5mm punch to create a hole directly in the center of the nose (Figure ) to create an opening for the Person Sensor module's camera. Then complete sewing pattern instruction **3** as directed.

3. SEW THE HEAD AND EARS

Pattern instructions **4–8** describe sewing the ears and head. We will modify these instructions to create channels connecting the head and ear interiors, leaving a pathway for wires and LED strings to be added later.

Follow instruction **4a** as written. In instruction **4b**, when basting the ears to the head front, do not sew all the way across the ear bottoms. Instead, baste just the portion where the ear is

Baste here only

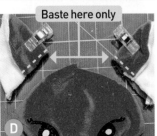

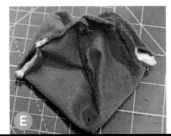

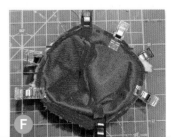

folded over on itself, as seen in the blue dotted lines in Figure  D.

Similarly, in instructions **4d** and **5**, baste the ears to the head front by sewing only the folded portion of the ears. Then separate the front and back sides in the unsewn ear half and baste just the light-colored ear fabric to the front of the face (Figure E).

Next, follow pattern instructions **6a–6c** as written to sew the head back pieces together, being sure to leave an opening as directed. Complete **7a** as written. In instruction **7b**, when sewing the head front and back together, stitch over the folded ear portions, but leave gaps where the open portions of the ears sit, as shown by the blue dashed lines in Figure F. Remove the assembled head from the sewing machine, then hand-stitch the dark ear fabric to the head back at the gaps in the machine stitching.

Once the head is assembled, with open channels from the head interior to the inside of each ear, invert the head as shown in instruction **7c**. Skip instruction **8** and do not stuff the head at this point. Set the head aside until the body is complete.

4. SEW THE BACK AND TAIL

Follow pattern instructions **9a–9c** and **10a–10b** to sew and turn the tail. Skip instruction 10c and do not stuff the tail.

Next, we will modify instruction **11** to leave an opening at the base of the tail to insert an LED string. Begin instruction **11a** by marking the body back pieces with tail placement lines. But instead of stitching through both layers of the tail, we will stitch each half of the tail base to a different body back piece, as shown in Figures G and H on the following page. It is easier to baste the small tail pieces by hand than by machine.

After basting the tail to both pieces which form Rosie's back, align and pin the two back fabric pieces together with right sides facing each other and the tail sandwiched between them. Sew along the line depicted in sewing instruction **11c**, *skipping over the tail opening*.

Once the body back and tail pieces are all sewn together, verify that you can still access the tail interior through the opening, then set the assembled body back aside.

Plushbot Menagerie

My experiments with electronics and plush patterns have resulted in this menagerie of charismatic companion bots with a range of capabilities:

1 **Rosie the Radiant Rodent** and her plush rat friends have illuminated ears and tails, sensors that "feel" touch and "see" faces, and servomotors that move their heads back and forth.

2 **Sluggo the Sloth** slowly turns his head from side to side, powered by a small stepper motor.

3 **Beacon the Soft Robot** has touch sensors in its hands, controlling LED patterns in its ears and antenna.

4 **Karma the Chameleon** drapes comfortably around her human companion's shoulders while the person detector in her head identifies faces, wagging her tail's 9g servo in response. Her head pivots up and down and side to side, powered by two standard servos on a gimbal, while touch sensors in her paws and head control colorful LED patterns in her ears and back.

All of these are adapted from sewing patterns by talented designer **Choly Knight,** whose creations showcase the bright, childlike design style known as *kawaii*, Japanese for cute or adorable. Choly is quite prolific, and publishes dozens of patterns, including those for the sloth and rat, for free on her website, **Sew Desu Ne?,** at cholyknight.com. Other patterns, such as the shoulder dragon and plush robot, can be purchased at her Etsy store (etsy.com/shop/CholyKnight). Her projects span a wide range of skill levels, and each tutorial contains comprehensive, detailed instructions that provide an engaging way to expand your sewing skillset.

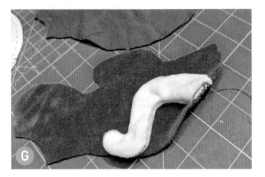

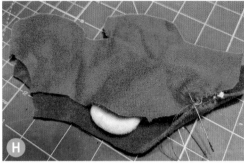

5. INSERT THE ZIPPER

Now we'll deviate from the pattern instructions to insert a zipper into the bottom fabric piece, to provide access to the finished bot's interior.

Using a Sharpie or fabric pen, draw the blue and orange lines shown in Figure ⓘ directly onto the wrong side of the quilting cotton piece. The long orange line should be centered inside the ½"-wide blue box, and the diagonal lines should branch off at 45° angles. The short ends of the box sit about ¾" from the nearest fabric edges.

Next, sandwich the quilting cotton and plush body bottom pieces with right sides together and pin or clip around the exterior edges to hold them in place. Take both body bottom pieces and place them in the sewing machine with the quilting cotton on top. Referencing Figure I, sew completely along the blue lines forming the box. Remove the fabric from the machine and take out the pins. With fabric or embroidery scissors, cut through both layers of fabric along the orange lines, being very careful to stop just short of the stitches which form the box (Figure ⓙ).

Push the quilting cotton through the newly cut opening so that the connected pieces now have their wrong sides together, then smooth the quilting cotton so it lines up with the plush fabric (Figure ⓚ). Gently iron the seams of the

rectangular opening to flatten them.

Center the nylon coil zipper behind the opening, with the zipper front visible from the plush fabric side, then pin the zipper in place (Figure ⓛ). Be sure that the zipper pull is inside the opening. Attach a zipper foot to your sewing machine and carefully sew around the entire rectangular opening, staying close to but just outside the seam, removing the pins as you go (Figure ⓜ). After sewing the zipper perimeter, trim any zipper ends that overhang the fabric.

6. SEW BODY AND ATTACH HEAD

We return to pattern instruction **12** to attach the body back to the body front. Be sure to leave the zipper open when sewing them together (Figure ⓝ). Do not invert the body yet.

Skip instruction **13** as we are not stuffing the body yet. Instruction **14** attaches head and body together, but we will skip it and use a different technique that takes advantage of the zipper.

Make sure the head is right side out and the body is inside out. Orient the head and body as shown in Figure ⓞ and insert the head into the body through the open zipper, settling it in place so that the neck openings in both the head and body align together, with the head opening inside the body opening. Secure both openings together

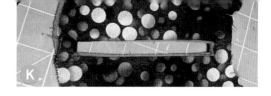

with clips or pins so that the front and back seams in the head and body align (Figure P). Carefully sew around the entire perimeter of the neck opening. Hand stitching is a bit easier than machine sewing here, but you can use either method to sew the head to the body (Figure Q).

Finally, pull the head back out through the zipper, pulling the body through behind it so that your plush rat is fully right side out. Check to make sure the ear and tail interiors are still accessible through the body cavity and the feet are fully extended. The plush rat shell is now completed and ready to house the electronics.

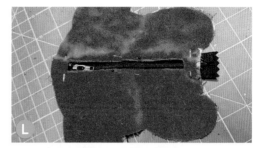

7. 3D PRINT THE FEATHER CASE

A custom 3D-printed electronics case will help protect soldered wires and provide a rigid base to hold the servomotor. Use the STL file *FeatherServoCase.stl* (makezine.com/go/cuddly-bot) to print the case and lid from PETG at 100% infill and 0.2mm layer height. While the case is printing, you may proceed with the next step to assemble the electronics.

8. SOLDER WIRES TO SENSORS AND LED STRINGS

We'll solder three different colors of soft 30AWG silicone wire to each of the two capacitive touch sensors and each of the three LED strings. Cut 4" lengths of different color wires for each touch sensor's VCC, GND, and Signal pins. Strip and solder one end of each wire to the sensor.

Cut three 7-pixel lengths of LED string, leaving as much on the input end as possible. Using fingernails or a snipping tool, separate the three input wires, then strip a short length of insulation from the ends. Cut three different color wires in 6" lengths for each LED string and strip the ends. With your soldering iron, tin the stripped portions of the LED strings and stranded wire ends.

Now solder and reinforce the LED string solder joints as shown from top to bottom in Figure R on the following page: Place a blob of solder on the soldering iron, and carefully melt the ends of the stranded wire to the LED string inputs. Slide narrow segments of shrink tube over each solder joint and heat the tube to shrink it. Once each solder joint is reinforced, slide an approximately 1" segment of transparent 8–10mm diameter

shrink tube over all three covered solder joints and the first LED in each string. Heat to shrink the transparent tube.

9. CONNECT CONTROLLER AND PERIPHERALS

Only the female half of the servo extension cable is needed. Cut the female connector from the cable along with about 2" of its wires. Solder its power and signal wires to the Feather's Bat and D13 pins, respectively, and leave the ground wire unsoldered for the moment.

The wired connections between the Feather RP2040 and its peripherals are shown in Figure ⓢ. The Feather's 3V and GND pins provide power for three LED strings and two touch sensors.

To simplify the connections, we'll solder the peripherals' power wires together in two bunches for power and ground, then connect each bunch to a single segment of solid core wire which then connects to a single Feather pin. Cut two 1" lengths of 22 AWG solid core wire and strip about 1cm of insulation from one end of each wire. Group the 30 AWG power wires from the capacitive touch sensors and LED strings together, stripping about 1cm of insulation from their ends. Twist the stripped ends together. Do the same with the ground wires from the touch sensors and LED strings, including the servo cable's ground wire in this bunch. When twisting the wires together, try to minimize tangling and knotting. Tin the twisted bunch of power wires, then solder it to the stripped end of one of the solid core wire pieces. Slide a piece of shrink tube over the joint and heat to shrink.

Repeat the same steps to connect the bunch of 30 AWG ground wires and the servo cable ground wire to the remaining piece of solid core wire. Finally, solder the free ends of the solid core wire pieces to the 3V and GND pins on the Feather 2040. At this point, the wiring should resemble Figure ⓣ.

Next, solder the free ends of the touch sensors' signal wires to Feather pins D24 and D25, and the LED strings' signal wires to Feather pins D10, D11, and D12.

Take the Person Sensor, and carefully use pliers to snap of the portion of the board labelled "break off" (Figure ⓤ). Connect the Stemma QT cable to the corresponding connectors on the Feather and the Person Sensor.

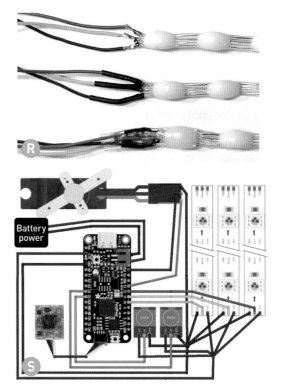

We'll coat the sensor boards with clear nail polish for an extra layer of protection and insulation. Take the touch sensors and paint a layer of polish over the entire side containing the tiny SMT components (Figure ⓥ). Let the polish dry for 10 minutes. On the opposite side of the touch sensors, paint a bit of polish over the solder pins only, avoiding the portion of the board with the word "touch."

Similarly, apply a coat of clear polish over the surface mount electronics on both sides of the Person Sensor, being careful not to paint the camera lens or the Stemma QT connector.

When the polish has dried thoroughly, insert the Feather into the 3D-printed enclosure, with the peripheral wires extending from the enclosure's side holes. Secure the lid to the base with the four 12mm M2 screws and nuts as shown in Figure ⓦ.

Attach the micro servo to the two vertical prongs on the 3D-printed case with the 8mm M2 screws and nuts as shown. Orient the servo so its output shaft is centered along the short end of the enclosure lid. Attach the single-ended horn to the shaft so that the horn points directly upward

in the middle of its rotation range, then screw the horn firmly onto the servo. Plug the servo connector into the female connector extending from the Feather.

10. PROVIDE POWER

Using a small Phillips screwdriver, insert the red and black wires on the JST cable into the +/– screw terminals on the barrel jack connector. Insert the USB cable into the female barrel jack and plug the other end into a 5V USB power source. Insert the JST connector into the battery connector on the Feather. The indicator light on the Feather will turn on to show that it's powered. Note that the LEDs and sensors will light up (Figure ⓧ) when either the Feather's USB-C port or its battery port are powered — but the servo will only move if power goes to the battery port.

11. CODE YOUR COMPANION

Code the Feather now to test all the electronics before installing them into the companion bot body. Install the latest version of CircuitPython on the Feather by following the instructions at learn.adafruit.com/welcome-to-circuitpython/installing-circuitpython. Connect the Feather to your computer where it should show up as a drive named *CIRCUITPY*. Then copy these CircuitPython libraries to the *lib* folder on the Feather:

- *adafruit_bus_device*
- *adafruit_fancyled*
- *adafruit_led_animation*
- *adafruit_motor*
- *adafruit_debouncer.mpy*
- *adafruit_ticks.mpy*
- *neopixel.mpy*

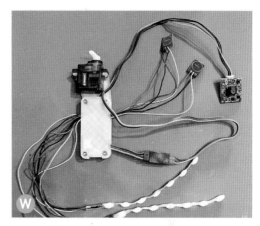

Copy the project code file, *code.py*, from makezine.com/go/cuddly-bot over to the main directory on the Feather and watch what happens to the attached peripherals. Comments in the code explain what the different sections do. Verify that your build has the following functionality:

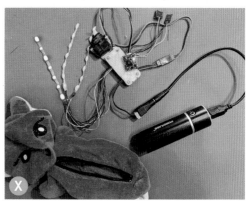

- The servo horn makes periodic small, random-appearing motions and returns to rest at center; you can adjust the horn, or adjust **rest_angle** in the code
- The tail LED strip runs continuous animations
- Contact with the paw touch sensor changes the

animation pattern on the tail LED strip
- Contact with the head touch sensor causes the two ear LED strips to display a purple animation and makes the head servo move vigorously back and forth for a few seconds
- The Person Sensor displays a green light next

to the lens whenever it detects a human face, and one of the two LED ear strips, determined by which side of the sensor your face is on, shows a blue animation.

12. SECURE SERVO TO THE HEAD

Once the electronics are working, they can be installed into the plush body, starting with the servo which moves Rosie's head. A piece of nylon zip tie attached along the center head seam provides leverage for the side-to-side motion.

Invert Rosie's head and push it out through the zipper opening. Thread a needle and hand-stitch the zip tie's "eye" near the point where all four head seams meet. Using a whipstitch, sew the zip tie along the head back center seam, stitching through the seam allowance and around the zip tie, until you reach the neck opening (Figure Y). Tie off and cut the thread there. With scissors or flush cutters, cut the protruding end of the zip tie about 1" beyond the neck opening.

Take a small piece of gaffer tape about ½" wide and 1½" long. Hold the servo horn against the cut end of the zip tie (Figure Z) and use the tape to bind them together tightly (Figure Aa).

13. INSERT ELECTRONICS AND STUFFING

Placing the electronics inside Rosie's small body can be tricky, requiring patience and perseverance to navigate the narrow openings.

First, place the LEDs in Rosie's ears. To facilitate this, first push the aperture leading into the ear toward the neck opening. Insert the free end of the LED string into the ear channel, then manipulate the string from the exterior of the ear. Push the leading LED around the outer curve of the ear, until the string forms an arc which follows the outer ear seam. Repeat with a second LED string in the other ear.

> **TIP:** If you inadvertently swap positions of the LED strips or touch sensors, you can fix the issue in the code, by changing which Feather pins control the misplaced peripherals.

Next, place the Person Sensor inside the head so that the camera lens protrudes through the hole in the nose, and the Stemma QT connector is oriented toward the top of the head. You may need to tug gently on the face to stretch the hole.

Now insert the Feather/servo assembly into the body so the servo horn sits just behind the neck opening, pointing up toward the top of Rosie's head. Add small bunches of stuffing to the head by compressing them and pushing them toward the rat's nose. Keep adding stuffing in small quantities, massaging it with your fingertips to eliminate bunching. When there is a moderate amount of stuffing inside the head, slide the head capacitive touch sensor through the neck opening so that it rests near the back top of the head, just to one side of the zip tie, with the touch sensor's flat side against the fabric. The stuffing will hold it in place. Keep adding stuffing until the head reaches the desired degree of firmness.

Once the head is stuffed, slide the paw touch sensor into one of the front paws, with the flat side touching the top of the paw. Place some stuffing inside the paw, underneath the sensor, to hold it in place, and then stuff the other three paws to match.

Slide the last LED string into the tail opening, then manipulate the tail from the outside to pull the leading LED through the channel to the tip of the tail. The tail should have enough structure to stand up on its own, but if you'd like to add some stuffing around the LED string, you can push tiny tufts gently into the tail using a chopstick.

All the electronics are now placed inside Rosie's body. Finish stuffing the body in the same manner as the head, until you are happy with the firmness.

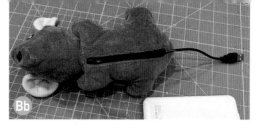

Try to keep the stuffing from settling between the Feather enclosure and the zipper so that it won't catch in the zipper. After stuffing the body, close the zipper, leaving the USB-A end of the power cable protruding outside the body (Figure Bb).

ROSIE RESPONDS!

Connect the power cable to a 5V source and watch Rosie come alive. Touching the sensors in her paw and head will cause the sensor light to illuminate through the fabric (Figures Cc and Dd). Touching the paw sensor will change the LED pattern in Rosie's tail, and patting her head will make her ears sparkle purple while her head bobs from side to side.

When the touch sensors are not triggered, Rosie's head will make small random motions from side to side, stopping whenever a face enters her field of view. Upon detecting a face, the green light on the Person Sensor will light up, and the ear on the same side as the nearest detected person will light up in blue (Figures Ee and Ff).

NEXT STEPS

The build may be finished, but you can still alter Rosie's personality. Edit her CircuitPython code to modify her responsiveness and behavior. For example:

- Try using Adafruit's Debouncer library to enable detection of long presses and double taps on the touch sensors.
- Currently Rosie reacts to only the nearest detected face, but the Person Sensor returns data on up to seven faces at a time. Rosie could react in different ways based on the size of the crowd around her.

Accessories offer a fun way to enhance — and wear — your companion bot. I sewed a spiked vest that snaps around Rosie's torso (Figure Gg), which provides both fashion and function: it can fasten around a shoulder strap or wristband, securing Rosie to my shoulder or arm while I walk around.

If you're in the mood for a bigger challenge, create an original bot based on a new plush pattern. Try adding extra motors to move its arms, legs, or tail, or look for new sensors to add functionality and responsiveness. Adafruit's large stock of sensors and their comprehensive Learn Guide library are good sources for new ideas.

MAKING NEW FRIENDS

The most surprising finding from my plush bot experiments has been the level of affection I feel toward my new creations. Even powered down, their presence on my desk while I write this article provides a comfortable camaraderie. Though I built them piece by piece and coded their every behavior, my companion bots exude personalities that exceed the sum of their parts. Constructing them has been an enjoyable, if unexpectedly literal, way to make new friends, and one can never have too many friends. ⊘

Personal Pixel Device

Build your own highly customizable pixel display with speaker, for a cool clock and much more

Written and photographed by Charlyn Gonda

CHARLYN GONDA is a coder by day, maker by night, from Alameda, California. She loves to create delightful, often glowy things that bring a bit of joy into the world.

Each of us has a different relationship to time, and that relationship tends to vary depending on our mood and what we're doing, among other factors. However, most clocks are built for passive observation, and don't capture our sense of style or how we personally interact with time. It's time we take matters into our own hands!

Let's build a highly customizable clock companion, that could be made to display whatever information matters to you while matching your unique aesthetic. I've made mine with an ambient sound visualizer, a simple focus timer, and a custom alarm that is only triggered at a specific time of day (to remind me to get to my ferry on time!). It's built with the Adafruit RP2040 Prop-Maker Feather, an all-in-one microcontroller that makes it easy to work with speakers, servos, LEDs, and more! The code that powers this device is written in CircuitPython for maximally easy code changes.

For the enclosure, I've upcycled a golden bottle cap that I've had for years, and the design of the case is heavily inspired by those beautiful Braun radios and phonographs from the 1950s and 60s. If you don't have access to a 3D printer or laser cutter, you can send the files out to a service to print/cut the parts for you. You can use any old knob, or maybe even model your own 3D-printed parts for this.

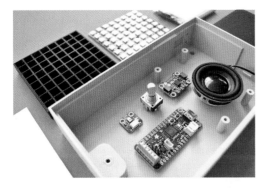

TIME REQUIRED: 2–3 Hours

DIFFICULTY: Moderate

COST: $150–$170

MATERIALS

» **NeoMatrix 8×8 NeoPixel RGB LED grid** Adafruit 1487, adafruit.com
» **RP2040 Prop-Maker Feather microcontroller** Adafruit 5768
» **Lithium ion battery, 3.7V 2000mAh** such as Adafruit 2011
» **Speaker, 40mm, 4Ω 3W** Adafruit 3968
» **Real time clock (RTC) module, PCF8523** Adafruit 5189
» **Rotary encoder, panel mount** Adafruit 377
» **Acrylic sheet, black LED transmissive, 3mm thick, 90×90mm** aka "black glass," from TAP Plastics, tapplastics.com
» **Acrylic sheet, opaque pearl, 3mm thick, 60×90mm** TAP Plastics
» **Stemma cables, QT to QT, 100mm (2)** Adafruit 4210, aka JST-SH or Qwiic cables
» **Silicone covered wires, stranded-core** about 2 meters each in 4 colors, Adafruit 1970
» **Flex PCB, half-size Perma-Proto Board (optional)** Adafruit 1518
» **Coin cell battery, CR1220** Adafruit 380, for the PCF8323
» **Microphone, PDM, with JST-SH connector (optional)** Adafruit 4346
» **Heat set insert, M3×4mm (optional)** Adafruit 4255
» **Leather strip, 1" wide, 4¾" length (optional)**
» **Machine screws, flat head, M2.5×6mm (4)**
» **Machine screw, M3×6mm (1)**
» **Foam tape, double sided**
» **Paper tape, double sided** aka Nitto tape
» **Kapton tape**

TOOLS

» **Soldering iron**
» **Lead-free solder**
» **Screwdrivers** to fit your machine screws
» **Flat head screwdriver, 1.8mm** for RP2040 Feather terminals
» **Wire strippers/pliers**
» **3D printer (optional)**
» **Laser cutter (optional)**

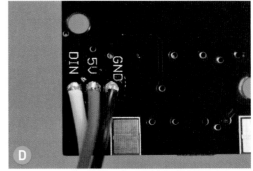

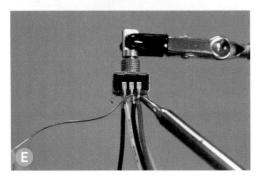

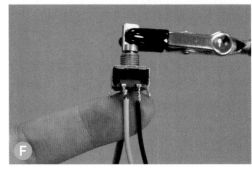

BUILD YOUR PERSONAL PIXEL DEVICE

1. PREP WIRES AND PARTS

1a. Fabricate case parts

First, we have to 3D print the main case body, the knob interface, and the grid frame, and then laser cut the front panels (Figure A). The 3D print requires no supports and should be easy to print.

Have fun with this — customize the material and color combos as you wish! I've shared a few versions of the files at github.com/chardane/personal-pixel-device, so you can print the case without the leather loop and print just one whole front panel instead of two pieces. You can also omit the microphone part if you don't need it.

1b. Attach leather loop

If you want to use the case version with the leather loop, you can attach the loop securely with a heat-set M3 insert (Figure B). Fold a 4.75" leather strip in half, poke a hole through, insert via the case slot, and attach with an M3×6mm screw (Figure C).

1c. Wire LED matrix and rotary encoder

Now we have to do a bit of soldering. For the LED matrix, solder a black wire to GND, a red wire to 5V and a yellow wire to DIN (Figure D).

The rotary encoder has three wires for the encoder and two for the button. On the side with three prongs, solder a black wire to the common ground in the middle, a yellow one to the left prong, and a red one to the right prong (Figure E).

On the other side, solder a black wire to either of the two prongs, and another color wire to the other prong; I use pink here (Figure F).

Now these parts are ready to be connected to the RP2040 Feather.

1d. Prep wires for optional microphone

If you choose to include the microphone, you'll have to prepare some wires. The PDM mic connects via a JST-SH connector on one side, but you still need to solder the wires to the appropriate pins on the Feather. So take one of the Stemma cables, cut off one QT connector, strip all the wires, and "tin" the exposed wires by dabbing a little bit of solder onto each (Figure G).

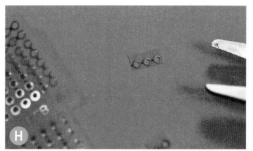

2. ASSEMBLE THE CIRCUIT

2a. Solder ground connections

The Feather only has one GND pin, but we need to connect three ground wires in total (for the encoder, the encoder button, and the mic). A common way to do this is to solder all three directly to one common wire and connect that to the GND pin. Another way is to cut a piece of flex PCB so you end up with a strip of three connected holes (Figure H). I used a bit of Kapton tape to secure the flex strip onto the Feather, and soldered one ground wire through both the GND pin and the flex strip. Then I soldered the two other ground wires to the remaining two holes on the flex strip (Figure I).

Note that this flex strip can break with repeated flexing — if this board stays inside the case it should be fine.

2b. Solder optional microphone wires

Take the remaining wires from the Stemma cable and solder the wires by color as shown in Figure J: blue wire to Feather pin A1, red wire to 3.3V, and yellow wire to TX.

2c. Solder encoder wires

Take the red wire from the encoder and solder it to pin 11, and solder the yellow wire to pin 10 (Figure K). These two wires should be from the three-prong side of the encoder. That was the last bit of soldering you had to do, hooray!

2d. Attach Stemma wires

Happily, the RP2040 Prop-Maker has a bunch of terminals so we can attach the rest of the wires without having to solder.

First, attach the soldered Stemma QT wires to the optional microphone

Then attach the other Stemma QT cable to the dedicated QT port in the middle of the board, and attach the other end to the PCF8523 real-time clock board (Figure **L**).

At this point, you can also insert a 1220 coin cell battery into the PCF8523 RTC board.

2e. Attach terminal block wires

Now connect the speaker and LED wires into the screw terminal block (Figure **M**), following the labels on the underside of the Feather board.

- Speaker **red** wire to +
- Speaker **black** wire to –
- LED grid **red** wire to 5V
- LED grid **black** wire to G
- LED yellow wire to Neo

Finally, screw in the last remaining wire from the encoder button into the terminal labeled Btn (shown in Figure **N** with a pink wire).

After all that, you can connect the battery to the battery terminal too!

3. TEST THE SOFTWARE

This is the perfect time to load our CircuitPython code and make sure we're good to go before we stuff everything into our beautiful case.

3a. Check connections

We're now at the point of the project that I like to call "wonderful wire chaos" (Figure **O**). You should take a moment to confirm that there are no loose or exposed wire ends, and that the screws on the terminal block are tight.

3b. Load code and test

CircuitPython makes it easy to put code into microcontrollers, because all you have to do is open up a text editor and edit your code — no special software required!

Connect your Feather to your computer via a good USB-C data cable. Visit Adafruit's site to learn more about CircuitPython and follow the

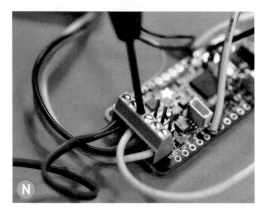

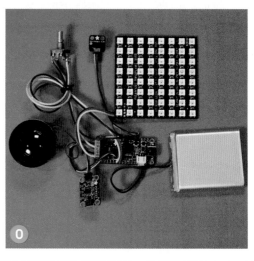

steps to make sure the version of CircuitPython running on your board is up to date (learn. adafruit.com/adafruit-rp2040-prop-maker-feather/circuitpython).

The Github repository contains all the code needed for this project. Go to github.com/chardane/personal-pixel-device and replace all the files in your Feather's *CIRCUITPY* drive with all the files from the *CIRCUITPY* folder on Github.

Once copied over, check to see that the code works by tilting the Feather perpendicular to the table (Figure P). You should see the LED matrix light up with some numbers scrolling by! It's probably displaying the wrong time, but you can follow the easy instructions in the code to set the correct time.

3c. Test encoder and optional microphone
Carefully tilt the Feather so that the USB-C side is facing down, which will turn on the Timer mode. Rotate the encoder to the right a few times and you should see the number of lights increase (Figure Q). Click the encoder once to make sure the button is also wired properly — that should cause the last pixel to blink.

If you've attached the microphone, position the Feather perpendicular again, as in step 3b — the clock should show up again. Click the encoder button once to enter Sound Visualization mode. Make some noise and confirm that the pixels light up and react (Figure R). If they do, that means the mic works, hooray!

4. FINAL ASSEMBLY
Now that we've confirmed our wiring is all correct and the code works, we can assemble everything into the case!

4a. Attach Feather and RTC
Use some double-sided foam tape to attach the Feather to the case, making sure to push the USB-C port into the appropriately sized hole in the case (Figure S).

Then, use some thinner Nitto tape on the back of the PCF8523 board and attach it just above the Feather (Figure T).

4b. Attach speaker and battery
Route the battery wires so that the battery can be

positioned and stuck into place above the speaker with some double-sided Nitto tape.

Position the speaker so that it ends up roughly behind the "speaker grill" holes on the front panel, and use double-sided Nitto tape to secure it in place (Figure **U**).

4c. Attach optional microphone
If you've chosen to include the microphone, there's a place in the corner of the case to screw in the microphone board using two M2.5×4mm screws (Figure **V**).

4d. Secure LED matrix
Now it's time to put the LED matrix on top of

both the Feather and the RTC board. Carefully arrange the wires so they don't end up on top of the green terminal block on the Feather, and position the matrix so that the wires are at the bottom left corner next to the speaker (Figure **W**). All you need are two M2.5 screws here, attached diagonally from each other. You can choose to use four screws but it is likely not necessary.

4e. Assemble front panel and grid frame
Attach the rotary encoder through the hole in the front panel, using the washer and hex nut that came with the encoder (Figure **X**). Then simply place the grid frame on top of the LED matrix — the "ear" piece should be positioned on top

(Figure **Y**). This frame isolates each LED pixel so their light doesn't blend together.

You should be able to simply press-fit the front panel into the case. If the tolerance on your print is quite loose, you might want to consider using a narrow piece of Nitto tape to help ensure that the panel press-fits tightly enough.

4f. Attach black LED diffuser and knob

Finally, press-fit the black LED acrylic diffuser panel on top of the LED grid, and attach your chosen knob (Figure **Z**). Using a bit of force on the knob should be OK here since we want a tight tolerance, but just be careful.

Congratulations, you've created a wonderful, personalized masterpiece of your own!

PIXEL POWER PERSONALIZED

Your Personal Pixel Device is ready to serve you with the following functions:

- **Clock mode:** displays the time as an animated crawl (Figure **Aa**)
- **Audio visualizer mode:** click the encoder button once to switch to this mode (Figure **Bb**)
- **Focus timer:** rotate your device 90 degrees so that the USB-C connector is face down. Turn the rotary encoder knob to the right to increase the time, 1 minute per pixel, and click it to start the timer. Click again to cancel (Figure **Cc**).
- **Alarm:** triggers an alarm at a specific time (set in the code), and plays a new LED animation and a sound (Figure **Dd**).

Since the device is quite modular, it should be easy enough to take it apart and add or remove components as your needs and desires change over time. This makes a solid foundation for some bountiful hacking!

- The 8×8 pixel grid can be quite versatile for many types of displays.
- There's likely enough room inside to add a Wi-Fi co-processor and make this display internet-connected!
- The RP2040 Prop-Maker Feather even has a built-in way of driving a servomotor, so you can potentially add some sort of kinetic component to your desktop companion.
- The powerful speaker here could play all sorts of music files.

Aa

Bb

Cc

Dd

- You might even consider using materials like ceramic or clay for the enclosure, to make this device truly yours.

The limit: the sky. The possibilities: endless. I hope you enjoyed making your own highly personalized pixel device! Please share your creation at @chardane (IG, X) or charlyn@leds. social (Mastodon). ⬤

Summoning Super-Size Spirits

Make a massive LED Ouija board and send spooky secret messages

Written by Lee Wilkins

LEE WILKINS is an artist, cyborg, technologist, and educator based in Montreal, Quebec, a board member of the Open Source Hardware Association, and the author of *Make:*'s column on technology and the body and how they intertwine. Follow them on Instagram @leeborg_

Jason Kapalka

Here's how I made a *giant* Ouija board, to summon giant spirits of course. I made this with Kyle Chisholm for Arcana, an occult-themed bar in Vancouver, British Columbia. The board can be remotely controlled by Wi-Fi through Adafruit.io to send spooky messages to bar patrons.

This was an interesting and fun request that I was very excited to build! The idea was to make a Ouija board as large as possible. For me, this meant as close to a full 4-by-8-foot sheet of plywood as possible, so there would be no seams.

Originally we discussed the idea of making a physically moving indicator ("planchette") for the Ouija board. After looking at a variety of similar projects that used XY stages, strings, or other contraptions, we decided that the upkeep on a long-term installation was simply too risky. Ultimately, we can agree that everybody loves LEDs! The features we wanted to integrate were that employees could send messages on the Ouija board, adjust speed and color, and schedule messages for later. The piece also had to be strong enough to be shipped across the country from Montreal to Vancouver, and to hang on the wall for a long time without risk of being damaged — so no exposed electronics.

One challenge for this project is that it had to be usable by employees every day at the bar. Usually I make art that I set up, and I am responsible for maintenance and taking care of weird quirks. But in this project, the experience had to be, if not seamless, then at least predictable in a way I could explain to the client. We chose to use Adafruit.io as our IoT platform for this, because it has a variety of features that *just work* out of the box. For example, we were able to create a simple dashboard to control the device remotely (Figure A), which was also secure in case somebody tried to hack it.

BUILD A BIG OL' OUIJA BOARD
1. DESIGN FILES AND MACHINING
The first step was to create design files for fabrication. I only have a 2×4 foot laser bed, but I knew this had to be machined on a single piece of wood that was nearly 4×8 feet to avoid any imperfections. In order to achieve the complex front panel design, I used a tape masking technique. I had the entire sheet covered in extra

TIME REQUIRED: Weeks!
DIFFICULTY: Intermediate
COST: $700–$1,000

MATERIALS
Please excuse my mixing units, I'm Canadian.
- » **Plywood, 4'×8' sheets: ¼" thick (1) and ⅛" thick (1)** for Ouija board and back panel, respectively
- » **Pine boards, 1×3** enough to frame your Ouija board
- » **Clear acrylic sheet, 3mm thick** for letters
- » **ESP32-compatible microcontroller dev board** We used the ESP32-S3-DevKitC-1.
- » **LEDs (2)** for indicators
- » **Power switch**
- » **Power supply, 12V, 3A**
- » **NeoPixel addressable RGB LEDs** or similar. We used about 140 total pixels.
- » **Project box** for the electronics
- » **Screws, nails, hot glue, jumper wires**

TOOLS
- » **Laser cutter and/or CNC router** If you don't have one, you can access one at a local makerspace or job it out.
- » **Soldering iron**
- » **3D printer**
- » **Wire cutters / strippers**
- » **Drill**
- » **Computer with Adafruit.io account**

wide masking tape before it went into the CNC. I chose a ⅛" bit, the smallest that could easily cut through the plywood. Even though I knew I would reinforce the panel, I still wanted it to be rigid enough to not bow over time, so I chose a ¼"

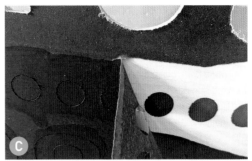

Lee Wilkens

plywood sheet with one sanded side.

The design was provided by Arcana, but I had to do some work on the Illustrator file to make sure there were no small details that would get chewed up by the machine: no sections of mask smaller than the size of the bit, and no sharp corners smaller than the radius of the bit. I cut the letters on the *interior* of the lines, but the letter cutouts (like the center of the Os) on the *exterior*, so that I could use the cutouts on the final design and they'd be proportional to the letters.

I made the letters all similar size and eliminated serifs or flourishes going past the main strokes, so I could create a uniform lighting system (a major issue in earlier attempts was the Q with a very long, curly tail). In the end, a bold, sans-serif font was the answer. I used red lines to indicate cut, and blue lines to indicate a light etching that just removed the tape (Figure B).

2. PAINTING

I first pulled off the mask areas I wanted to be painted. I used a dark brown milk paint so the color wouldn't be solid black but would look a bit more rustic. I started by painting the edge of the board. When masking, it's important to make sure the edges of your tape are solidly adhered to the base, otherwise your lines will end up fuzzy;

using a plywood that had a smooth, sanded front was critical for good adhesion (Figures C and D). Still, after machining, some of the smaller pieces of tape needed manual fixing.

Much to the confusion of my neighbors, I then took the board outside to paint the whole thing with a wood stain. I wanted to bring out the grain of the wood and make it feel old. I chose a stain that I could apply over the brown paint without significantly affecting the color, so it ended up being pretty light, but I think it did a lot for the aesthetic (Figure E).

3. EDGE-LIT LETTERS

Originally I imagined placing LEDs behind each letter and using about 4" of space to diffuse the light, but I couldn't design a good rig to secure it to the front panel and also leave the back clear to be hung on the wall. I wanted it to be as bright as possible, so I thought direct light would work best. But after experimenting with a variety of techniques, I settled on ***edge lighting***. This technique involves LEDs shining sideways into the edge of a panel of plastic, typically acrylic, with laser-etched designs on the surface, where they catch the light and illuminate. Perfecting the technique took some experimenting; I had previously used it in larger designs with a lot more freedom in terms of space and color.

One thing I wanted was for the letters to be dark when they weren't lit up, which would have been easy with back lighting. However, the etched area of the acrylic panel looks white. I tried painting the back of the panels, but that stopped the light from bouncing around inside the acrylic and dimmed the effect significantly. I also tried using frosted acrylic in order to help the diffusion, but that blocked a lot of light too, so clear acrylic is really key! In the end, I painted the interior of the entire Ouija board black and that worked.

To make the letters, I laser-cut individual plates for each. I left a 3" space between the bottom edge, where the light would shine in, and the start of the letter, to let the light diffuse and avoid a sharp line. I etched only the outline of the letters with the laser and then, using the protective acrylic coating as a mask (Figure ), hand-sanded each letter (Figure ). This was to avoid any patterns from laser etching that sometimes occur, and also to save on laser time!

4. LIGHTING SYSTEM ASSEMBLY

To hold the LEDs to the edge of the acrylic, I 3D printed custom mounts (Figure ). I made two parts to ensure they were securely held. The first part was a track for the LEDs to be threaded through, and to hold the acrylic panel. The trick is for the 3mm acrylic to be held exactly in the center over the LEDs so all the light goes through the panel; the closer it is held, the better the effect. I included holes for mounting screws so the track would be easy to attach to the front panel. The piece had to be thick enough to not crack, but also fast to print because I needed so many! I ended up with a simple design that held the acrylic tightly, but left room for the LEDs to be easily threaded through (Figure ).

I also made a standoff to hold the top of the panel in place (Figure). While very small, this piece was critical in stabilizing the letters; the acrylic can't lay flat against the front panel because it has to be centered over the LED strip. I printed as many of these as I could and secured them with a screw and hot glue (Figure ). I kept them small because space was getting tight between the rows of LEDs!

I placed all the letter panels over their

corresponding holes, and used small screws to attach each lighting track. Then I threaded the strip of WS2812 LEDs through each track, careful that each letter had three lights each so I could address them all individually. I can't stress enough how important it is to have some wiggle room in your LED track when you need the same string of lights to go through 32 different pieces. I also watched the direction of the LEDs to make sure all the arrows were consistent, so I wasn't left with any surprises when I got to the code.

I then glued down the standoffs and secured the panels in their sockets, making sure they were as close to the lights as possible. I also added some bracing and framing with 1×3 wood boards, and a back panel to make sure it stayed in shape (Figure L). Finally, I glued on the letter interiors to make it look finished (Figure M).

5. LEDS AND CODE

We used an ESP32-S3-DevKitC-1 (Figure N) as the microcontroller for this project because it was easy to set up and connect to the internet, and easy to use MQTT to control — and because it has a dual core.

Adafruit.io uses the MQTT protocol to communicate between the board and the application (learn.adafruit.com/welcome-to-adafruit-io/adafruit-io-mqtt-api). In simple terms, this works by subscribing to *topics* which each device is listening to and sending messages to. Our Ouija board uses both cores in the ESP32-S1: one to listen to the MQTT topic and decode the message on every loop, and the other to constantly display the most recent message posted to the topic. A *semaphore* is used to control access to the shared data between the cores, kind of like passing a ball between the threads so that data being used by one thread isn't interfered with by the other. FreeRTOS has a set of tools and functions that facilitate *inter-thread communication (ITC)*.

I'm going to share a few highlights that made this project interesting, including how multithreading works and how to decode ASCII values into LED indexes. Here we are creating a *mutex* with type **SemaphoreHandle_t**, as well as a Boolean to check if there is a new phrase, and a string to hold the phrase:

```
SemaphoreHandle_t semaphore =
xSemaphoreCreateMutex();
    bool new_phrase;
    String phrase;
```

On one core, MQTT is constantly checking the topic for new input. If the semaphore is available, it is taken so that the information isn't interfered with by the other core while it is working, by using **xSemaphoreTake()**. If a new phrase has been received, the animation is reset, **new_data**

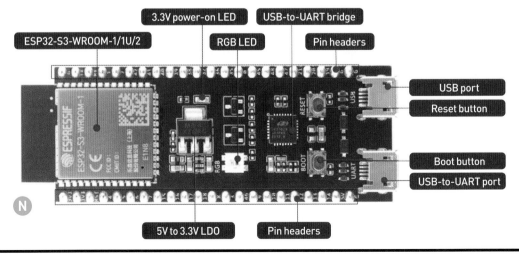

ESP32-S3-WROOM-1/1U/2 · 3.3V power-on LED · USB-to-UART bridge · RGB LED · Pin headers · USB port · Reset button · Boot button · USB-to-UART port · 5V to 3.3V LDO · Pin headers

is false, and we return the semaphore using **xSemaphoreGive()**. There are lots of ways to do this; I like CircuitState's tutorial at makezine.com/go/freertos-multitask, and I encourage you to read more about FreeRTOS.

```
        if (xSemaphoreTake(semaphore, 5)
== pdTRUE)
{
        if (new_phrase) {
            // reset running animation
with the new message
            resetAnimation(phrase);
            // unset "new_data" Boolean
            new_data = false;
        }
        // release mutex so MQTT
thread can update the data
        xSemaphoreGive(semaphore);
    }
```

To decode the input from Adafruit.io into positions on an LED string, we used a 2D array. This array stores the numerical position of each LED. For example, each letter has about three lights, but some have two, like I which is narrow, or four, like W which is wider. It's tedious, but each of them is mapped out.

```
int g_letters[OUIJA_NUM_LETTERS][2] = {
  { 124, 126 }, // 0 is a
  { 120, 122 }, // 1 is b
  { 116, 118 }, // 2 is c
  { 112, 114 }, // 3 is d
```

We're then able to use the ASCII character sent by the string to determine the position in the array. For example, ASCII 97 to 122 are the lowercase alphabet from A to Z. If we get 97, we know it's an alphabetical input, so we can subtract 97 to get the position of the LEDs in the array: if we get **a**, 97–97=0, and **a** is the 0th position in the array.

```
if (ascii_character >= 97 && ascii_character <=
122) {
    // numbers 97 to 122 are a to z
    return (ascii_character - 97);
```

The same is done for numbers, as well as Yes/No, which are indicated by capital **Y** or **N** letters:

```
[...]if (ascii_character == 89) {
    // 89 is Y
    return 36;
} else if (ascii_character == 78) {
    // 78 is N
    return 37;
```

PARANORMAL PORTAL PROGRAMMING

If you want to see a full example of how the code went together, you can check it out on my GitHub at github.com/LeeCyborg/GiantOuija. There are a few great pieces in there, including using Perlin noise to make a subtle background animation when letters aren't being shown.

I'm really happy with how this project turned out, and I hope to see more giant Ouija boards out there! ◑

Hamming It Up

Turn a can of ham into a functional ham radio antenna

Written and photographed by Ben Eadie, VE6SFX

BEN EADIE, aka Ham Radio Rookie, is a movie prop maker, practical special effects designer, inventor, nerd, and former aerospace engineer. You can see his work in films like *Star Trek Beyond*, *Predator*, and *Ghostbusters: Afterlife*.

Yes, you read that right. A tin of ham can indeed be transformed into a ham radio antenna. In the DIY world, where creativity intersects with utility, the question isn't "Can it be done?" but "How can we make it happen?" This ethos is at the heart of my latest project in my ham radio journey: I am showcasing my adventures as a newly minted ham radio operator less than a year into having my license, on my YouTube channel Ham Radio Rookie, youtube.com/@VE6SFX_HamRadioRookie.

"WILL IT HAM?"

This project started as a tongue-in-cheek dare from my friend Marshall. At a gathering, he asked if I could "turn anything into an antenna," referencing my videos where I had already transformed duct tape into a J-pole style antenna for a handset. In my typical bravado, I said, "Damn straight, I can!" Before I could process the absurdity of my claim, Marshall challenged me to turn a can of ham into an antenna. The gauntlet had been thrown, and no self-respecting maker backs down from such a dare. Thus, "Will It Ham?" was born — a video series where I turn everyday items into functional ham radio antennas, starting with a humble can of ham.

THE HAM CAN CHALLENGE

The challenge was straightforward yet bizarre: repurpose a can of ham into a functioning 70cm-band ham radio antenna. The first step was obvious — emptying the can of its original contents. We didn't need a "meaty" signal, after all. (Although ever since the video's release, I've been planning to try it with the ham still inside. Why? Well, if I push enough power into the can, I might be able to cook the ham by talking to it. Literally.)

A LITTLE HISTORY

I've been researching a design called a **slot antenna**, commonly used for TV broadcast antennas. It's essentially a slot cut out of a piece of metal. Here's where a bit of history and science comes into play: antennas with vertical orientation are well-received by other vertical antennas but not as well by horizontal ones, and vice versa. When TV broadcasts were introduced,

TIME REQUIRED: 2–3 Hours
DIFFICULTY: Easy
COST: $20–$50

MATERIALS
» **Canned ham** Amazon B00LZD1F9G
» **Copper foil tape, ¼" wide** Amazon B01MR5DSCM
» **BNC coax cable** Amazon B0BG2C4L7Q
» **BNC female connector**
» **BNC to SMA adapter** Amazon B08QHRHQRZ
» **Ring terminal crimps** Amazon B01D848CHM

TOOLS
» **Electrical pliers/crimpers** such as Amazon B08BX9RTPX and B01KN9YUSU
» **Handheld transceiver (HT)** I recommend Tidradio H8, tidradio.com. Use referral code "BEN" and you will get 15% off.
» **3D printer** I recommend the Creality Ender 3 V3 SE.
» **Antenna analyzer (optional)**

they interfered with standard FM radio communications due to their **vertical polarization**. To minimize this interference, TV broadcasts were mandated to use **horizontal polarization**. And the horizontal footprint of a normal antenna this big was a problem, because they are long and difficult to rig up in the air without being damaged by weather.

But here's the twist: slot antennas, when vertically oriented, have horizontal polarization! This solved the TV problem. And if we turn this slot on its side for smaller antennas, they become vertically polarized. For the 70-centimeter band frequencies, most if not all communications are vertically polarized.

One more thing about slot antennas is that

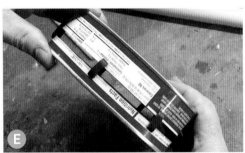

they don't need to be a straight slot, they can be bent or curved and still operate. So the shape of the ham can is not an issue here; the length of the slot is far more important. For more on this, check out *Slot Antennas for Ham Radio: The Forgotten Antenna* by John Fortune, W6NBC, available on Amazon.

THE BUILD PROCESS

After cleaning the can (Figure A), I used 3D modeling to create spacers for optimal spacing between the can halves (Figure B), then printed them out (printables.com/model/643707-hamtenna) (Figures C, D, and E).

A PVC pipe, attached with hot glue, formed the base (Figures F and G). Careful sanding (Figure H) prepared the electrical attachment points for the two halves, to bridge the slot (Figure I), and to connect a short coax cable terminating in a BNC female connector (Figure J).

Ideally, the perimeter of both sides of the can combined would add up to 70cm. As it turned out, this perimeter was almost exactly 70 centimeters!

TRIAL AND ERROR

Any maker project involves trial and error, and this was no exception. The *feed point* of the slot antenna is critical. It determines whether the transmitted energy radiates effectively or, in the worst case, sends it all back into the radio, releasing the dreaded magic smoke. Finding the

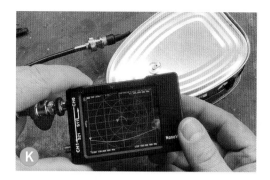

perfect feed point involved a mix of calculations and intuition. A good starting point is about 1/3 the distance from one end of the slot. Using an *antenna analyzer* (Figure K), you can fine-tune the position to achieve the ideal resonance.

To my complete amazement, not only did the antenna work, it worked quite well! You can see the surprising results at youtu.be/xQxB8ci9R_Q.

BEYOND THE BUILD

This project was more than a technical exercise; it was a celebration of the maker spirit, demonstrating that with a bit of creativity, even the most mundane items can be transformed into something functional and new.

The success of the ham can antenna is just the beginning. I'm calling on the maker community to suggest what should be "hammed" next. The weirder, the better. A tinfoil hat antenna? Did it! See it at youtu.be/6y0lfBJQ5ql and share your ideas in the comments! ◑

Watchful AI Owl

Make a simple animatronic robot, with AI that knows how to follow your face

Written by Reade Richard,
Brenda Shivanandan,
Andy Forest, and
Denzel Edwards

Whether you want to make a robot that can navigate your room, a creature that responds to your movements, or a gadget that can recognize your voice, you can build it with household materials and an inexpensive micro:bit mini computer — and then train an external AI to give your project extra smarts and capabilities, enabling it to react to the world around it.

YOUR FIRST ANIMATRONIC

Have you ever noticed how some robots look and act in realistic ways, like people or animals? You can see these used often in movies and at theme parks. These are called *animatronics* — or more simply, mechanical puppets controlled by lots of electronics (and even more code). While we may not be able to build something as sophisticated as a theme park puppet here, we'll show you the basics of having your puppet look around and flap its wings at your command.

Plenty of unique animal creations can be brought to life by mixing mechanical linkages and AI. Check out this owl we made! We used a simple three-bar linkage mechanism to make the wings flap, and added a second servo to turn its head to mimic scanning the environment for

TIME REQUIRED: A Weekend
DIFFICULTY: Easy/Moderate
COST: $28-$35

MATERIALS
» **BBC micro:bit mini computer** with USB cable
» **Micro servomotors, SG90 (2)**
» **Alligator-to-pin wires (6)**
» **Thick corrugated cardboard**
» **Thin cardboard** like a cereal box
» **Straws**

TOOLS
» **Computer with internet access and webcam**
» **Scissors**
» **Glue**
» **Tape**
» **Hole punch**
» **Hot glue gun**
» **X-Acto knife**
» **Servo tester (optional)**

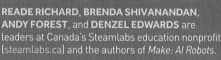

READE RICHARD, BRENDA SHIVANANDAN, ANDY FOREST, and **DENZEL EDWARDS** are leaders at Canada's Steamlabs education nonprofit (steamlabs.ca) and the authors of *Make: AI Robots*.

predators and prey.

You can even train this owl to make its eyes follow you around the room, by training an AI image-recognition model with three classes: one of you on the right side of the screen, one in the middle, and a final one on the left. Then you'll need a bit of code that will help you turn those servos in the right directions when the AI detects a change.

Add some more code to flap the servo wings every now and then, and you'll have a curious owl that scans a room and can react to curious visitors!

BUILD YOUR FACE-FOLLOWING AI OWL

BASIC BUILD
Build your owl's body and head out of corrugated cardboard or a similar material. If you like, you can use our patterns at makeairobots.com/chapter8. We've given you a Basic Head and Basic Body if you need a starting point for your puppet pal, but don't let us get in the way of your creativity — make something unusual or more complex if you're ambitious. Think of what you have in the recycling bin; milk cartons are a great replacement for a body, and yogurt containers could be used to make a head.

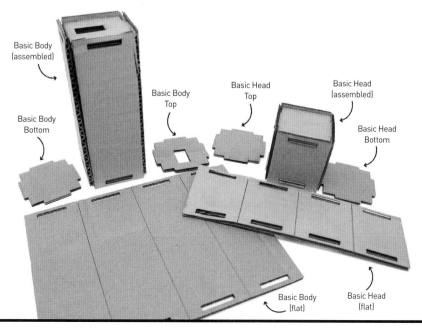

Basic Body (assembled)

Basic Body Bottom

Basic Body Top

Basic Head Top

Basic Head (assembled)

Basic Head Bottom

Basic Body (flat)

Basic Head (flat)

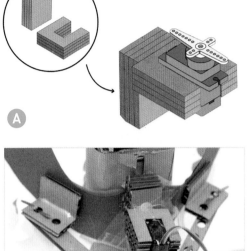

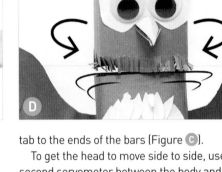

Brenda Shivanandan

Linkages are a great tool to help automate the flapping wings. Make sure you design the wings out of a material that can easily bend. The servo pushes and pulls on the wings to move them back and forth, so consider using thick construction paper or thin cardboard. You'll use a three-bar linkage connected to the servo to help the wings move.

Start by mounting the servo to the center of your owl's back — it doesn't matter if it's facing up or down. We connected ours by creating a small U-shaped structure to hold the servo and gluing that piece to another small group of rectangles to give it some extra space, so the linkage doesn't bump into your owl (Figure A).

Glue a short linkage bar to the servo horn; a five-hole linkage bar should be long enough to do the trick. The other two linkage bars need to be long enough so they extend out from this bar about 1 or 2 inches past the servo structure. Connect the bars with brass fasteners or a straw (Figure B).

Build two small cardboard tabs, and punch a hole in each tab before securing one onto each wing. Use a brass fastener or straw to attach the

tab to the ends of the bars (Figure C).

To get the head to move side to side, use a second servomotor between the body and head to control the head movements (Figure D).

AI TRAINING

Google's Teachable Machine (Figure E) allows anyone to create an AI prediction model to recognize images (or even sounds). Training the AI model for this puppet, you'll specify three different classes of images you want the AI model to be able to recognize. You can refer back to "Anyone Can Use AI," *Make:* Volume 84, makezine. com/go/ai-and-microbit, for general instructions, but the Teachable Machine website will walk you through it all.

Because we want the owl to spy on us as we move to either side of the room, you'll need to actually move to that side of the room when taking your reference photos. You may need a second person to take the pictures as you move around the room if you're far from the camera controls. Remember to name your classes something clear, like Left, Middle, and Right.

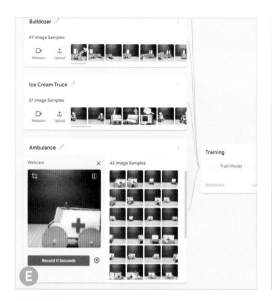

PROJECT WEBSITES

- **MAKE: AI ROBOTS:** Code, links, and instructions for adding AI image recognition to your micro:bit using your webcam. Print-and-cut PDF for Owl mechanism. Code and electronics troubleshooting: makeairobots.com

- **MAKECODE:** Program your micro:bit: makecode.microbit.org

- **TEACHABLE MACHINE:** Train your AI: teachablemachine.withgoogle.com

CODING

Hook the signal wire (orange) from the owl's head servo to pin 0 on the micro:bit, and the signal wire from the wing servo to pin 1 (Figure F). Hook both servo power wires (red) to 3V, and both ground wires (brown) to GND.

Make a new project at the Make: AI Robots website (makeairobots.com). Follow the steps there to connect to Teachable Machine and train your AI, then open up a coding window to program your micro:bit, and finally connect your micro:bit to your AI!

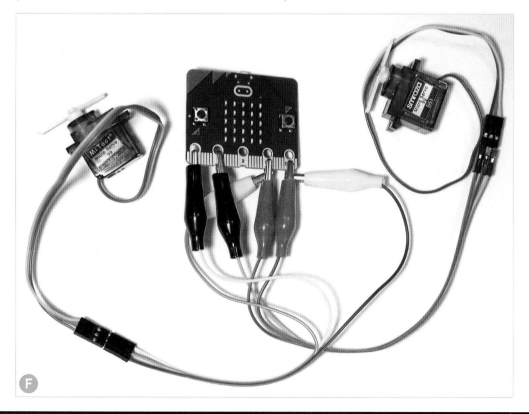

First let's set up the code to control the head. (If you're having trouble, the finished code for this project is available at makeairobots.com/chapter8 as "Animatronic Animals Makecode.")

1. Start with two **Set Servo (P0) Angle to (90)** blocks in the **On Start** block, setting the first to **(P0)** and the second to **(P1)**, but with both angles set to 90° for the owl's starting position (Figure **G**).

2. Set up the classes in the **If** statement to match your AI model: Left Side, Middle, and Right Side.
   Place one **Set Servo (P0) Angle to (90)** block in each opening, and use 45° for Left Side, 90° for Middle, and 135° for Right Side (Figure **H**). Now when your AI webcam is running, your watchful owl will turn its head to follow you around the room!

The code to flap the wings doesn't need to be controlled by the AI, so take an **On Button (A) Pressed** block from the Input drawer and place it next to all the other code.
   We're trying to recreate realistic wing movement with these servos, so think about what that looks like: Have you ever seen a bird spread its wings and flap them a few times in a seemingly random way? That's what we want to do here.

3. Drag a **Repeat (4) Times Do** block from the Loop drawer and place it inside the **On Button (A) Pressed block**, and set it to repeat two times. Then drag a **Set Servo (P0) Angle to (90)** block inside the loop and set it to **P1** so it controls the wing servo (Figure **I**).

4. Since real birds are unpredictable, the wing movements of this owl should reflect that; set this servo to turn to a random number.
   Open the Math drawer and take a **Pick Random (0) to (0)** bubble and place it into the degrees slot in the **Set Servo** block. Set the range to be between 0 and 80, which will be enough to push the wings forward (Figure **J**).

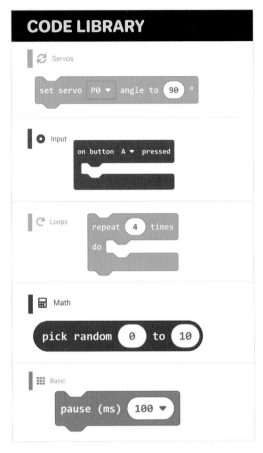

CODE LIBRARY

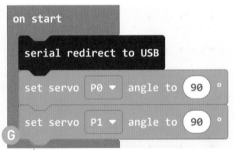

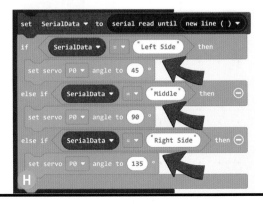

5. Add a **Pause (ms) (100)** block set to 250, which will provide time for the servo to move before moving again. Right-click on the **Set Servo** block you just made, and select Duplicate from the menu to make a copy of this code with the random number bubble already in it (Figure 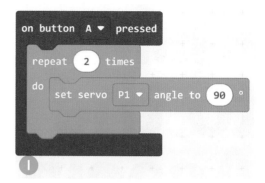). (This saves us from having to sort through multiple drawers again.)

6. Place this new block just below the **Pause (ms) (100)** block you made, and set the random range of this block as 95 to 180.

The first 0 to 80 range will make the wings move a little bit, then the 95 to 180 range will make the wings move much farther. Add another **Pause (ms) (100)** block set to 250 at the bottom so the second movement has time before the loop repeats.

Finally, place a third **Set Servo (P0) Angle to (90)** block after the loop and set it to **P1** so the wings return to their starting position (Figure 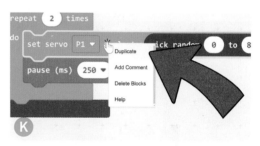).

Now whenever you press the A button on your micro:bit, your watchful owl will have a new pattern of flaps to show off! ◗

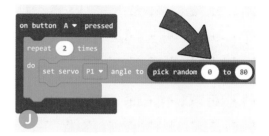

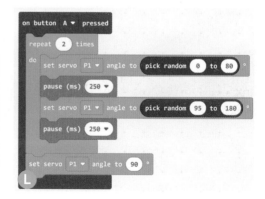

This project was excerpted from the new book *Make: AI Robots,* available at the Maker Shed (makershed.com) and fine booksellers.

Rising to the Acacia

Target hack: Turn a cutting board into a gorgeous end table in minutes

TIME REQUIRED: 30 Minutes

DIFFICULTY: Easy

COST: $20–$60 Per Table

MATERIALS
- » **End table with approx. 12"×15" top for frame,** such as Target 53107499 or Amazon B0B58PQX2S
- » **Cutting board, approx. 12"×15" or bigger** such as Target 87712370
- » **Wood screws, 1" (4)**
- » **Washers (4)**
- » **Wax, oil, or polyurethane finish**

TOOLS
- » **Drill**
- » **Punch or awl**
- » **Screwdriver**
- » **Paintbrush**
- » **Very fine sandpaper, 220 grit**

Written and photographed by David Battino

DAVID BATTINO (batmosphere.com) is the co-author of *The Art of Digital Music* and writes the popular "Synth Hacks" column for *Waveform*. With his wife, Hazuki, he also self-publishes Japanese storytelling books, which they have performed at four Maker Faires.

Stroll into any Target and you'll find beautiful wooden cutting boards with a tragic future: Most buyers will slash at them with knives, splatter them with food juices, and then toss the scarred wood in the trash. But these boards are a great source of quality wood for DIY projects. In just a few minutes, I replaced the plasticky tops of two end tables with gorgeous acacia hardwood. The hardest part was waiting for the finish to dry.

1. CHOOSE TABLES AND BOARDS
I upgraded a pair of C-shaped end tables that I got from Amazon. At Target, I selected cutting boards that were slightly larger than the original tabletop and had an interesting slanted edge (Figure Ⓐ). Even unfinished, the acacia looks way better than the fake wood-grain original.

2. MARK AND DRILL
Turning each frame upside down, I laid it on the bottom of a cutting board and marked the screw holes with a punch. I then drilled pilot holes for the wood screws (Figure Ⓑ). I could have extracted the threaded inserts from the old tops and used the provided machine screws, but because I planned to attach the frame to hardwood instead of crumbly particle board, I went with wood screws instead (Figure Ⓒ). I added metal washers for strength.

3. FINISH AND ATTACH
Before attaching the new tops, I sanded them and applied two coats of Danish oil and three of polyurethane (Figure Ⓓ), sanding lightly between each coat.

The C-shaped steel frames slide nicely under my couch. Next, I'll turn the original tops into garage shelves.

BOARDS GALORE
I used acacia wood for my project, but there are many other options, from affordable bamboo to exotic hardwoods, in endless interesting shapes. Cutting boards are a fantastic DIY resource. Even those slashed and dirty ones: Just add legs and turn them into soldering risers (Figure Ⓔ). The juice groove keeps electronic components from rolling over the edge. ⊘

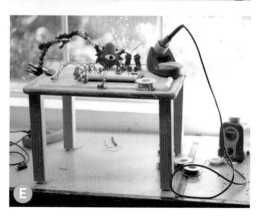

Walk This Way

Mobility walkers can be expensive. Build your own personalized, strength-tested version for less

Written by Joshua Pearce

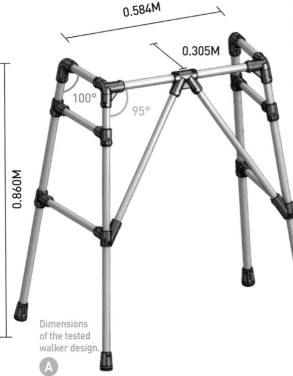

0.584M

0.305M

100°

95°

0.860M

Dimensions of the tested walker design.

B Testing a distributed load that replicates the span of hands on the walker, by crushing it in a giant press.

Walkers, a common adaptive aid for the mobility-disabled, are often not accessible or affordable across the global market. Private insurance and even countries with universal health coverage may not cover the cost of a walker, which can cost $100 or more. My colleagues Anita So, Jacob Reeves, and I at the University of Western Ontario were determined to learn if mobility aid costs could be reduced.

MOBILE DEVICE DEVELOPERS

After interviewing people who use walkers and examining commercially available models, our team developed a low-cost static walker, aka walking frame, by using freely available Onshape CAD software, simple hand tools, a 3D printer, and wooden dowels. Though strength will vary depending on the wood type chosen (e.g., basswood, beech, maple, oak, pine, etc.), we built it from relatively lower-strength basswood to assess our design conservatively. We tested several iterations with a constant focus on safety and stability, until finalizing this free and open-source design that's easy to build, adapt, and customize.

This walker's stability and rigidity are achieved by three main features: an A-frame design, triangular bracing that angles inward from the front legs and meets centrally on the top bar, and two horizontal braces on each side (Figure A). The wooden dowel frame parts are connected by tough PETG plastic connectors.

The three-piece foot design connects a rubbery TPU base to the dowels via a press-fit PETG body and #6 flat-head screws. These dual-material feet provide greater friction to improve the walker's stability on smooth surfaces such as tiled floors. Similarly, a convex cylindrical TPU handle can be press-fit on the walker's top sides to improve grip and comfort.

To ensure these walkers could safely withstand regular use, the team rigorously tested them to exceed the "static strength of walking frame" described in ISO 11199-1:2021, a standard that requires walkers to bear a purely vertical load of 1,500N — about 337lbs or 153kg — without cracking or breaking for a duration of 2 to 5 seconds. To study failure modes and locations, we exceeded this standard by testing the walkers to failure in a hydraulic press (Figure **B**).

Lateral loading was also considered. Initially, we played an energetic game of tug-of-war; controlled studies with a second hydraulic pump followed. You can read testing details and data at doi.org/10.3390/inventions8030079. Western did all the necessary testing for this design; you won't need to replicate it as long as you follow the manufacturing instructions.

These walkers are designed to be customized. Not only do you size it for a specific person, but you should also choose the user's favorite filament color!

We're honored that our low-cost walker is included in the Project Library at Open Source Medical Supplies (opensourcemedicalsupplies.org), whose Victoria Jaqua and Christina Cole helped prepare this article for Make:.

TIME REQUIRED: A Weekend
DIFFICULTY: Easy
COST: $50–$70

MATERIALS
» **Wood dowels: 19.19mm (0.755")**
diameter, 4' long (2) We used basswood, but you can use something stronger if you choose. Total length used: 2.27m (89").
» **Wood dowels: 22.4mm (0.88") diameter, 4' long (5)** Total length used: 4.12m (162").
» **3D-printed parts, in PETG**
See Table 1 for listing.
» **3D-printed parts, in TPU 85A** such as NinjaFlex, for foot cushion, washer, and handle grip only. See Table 1 for listing.
» **Wood screws, flat head, #6×½" (82)**

TOOLS
» **3D printer, FFF type**
» **Wood saw**
» **Screwdriver**
» **Drill with driver bit (optional)** to match your screws; very helpful

JOSHUA PEARCE runs the Free and Appropriate Sustainability Technology (FAST) lab at Western University in Canada.
appropedia.org/FAST

BUILD YOUR DIY MOBILITY WALKER
1. PRINT THE PARTS
Download the STL files from our page at Open Science Framework (osf.io/v3njw) then print the parts. Use NinjaFlex TPU filament for the foot cushion, washer, and handle grip, and PETG for the rest. We used an open-source RepRap-class 3D printer, but many options exist (e.g., Prusa i3, Lulzbot Taz, etc.). **Table 1** lists the names and quantities of the 3D-printed parts, and **Table 2** (on the following page) provides the slicing parameters for the PETG and TPU components.

Due to the intricate shapes of the parts, each was oriented strategically on the print bed to avoid aligning the print layers with the anticipated fracture failure planes (Figure **C** on the following page). A brim of 0.5mm is recommended for parts with minimal surface contact with the bed. Support is required for both the Angled Mid

TABLE 1. 3D-printed parts.

Name	# of Parts
Ang 3 Connector [L]	1
Ang 3 Connector [R]	1
Ang 2 Connector [L]	1
Ang 2 Connector [R]	1
Ang Mid Support [L]	1
Ang Mid Support [R]	1
Middle Support	1
Ang Side Support [O]	3
Ang Side Support [□]	3
Foot Body	4
Foot Cushion	4
Foot Washer	4
Handle Grip	2

Supports and the Foot Body components due to extreme overhangs; see Figure Ⓓ for parts with support locations and brim. Finally, the handle was printed in a vertical orientation (Figure Ⓔ).

2. MEASURE FOR THE USER AND CUT THE WOOD

To ensure the walker properly fits the user, you'll take measurements and adjust them with the calculation procedures detailed below. The quantity of each dowel part is specified in **Table 3.**

2a. Height of walker: Measure from the ground to the crease of the user's wrist while in an upright position with arms relaxed on the sides and wearing shoes.

2b. Length of Leg dowels: Take the height of the walker established above and divide by cos(10°). Subtract 15mm for the thickness of the foot cushion and 30.75mm for the top 3-dowel joint. The final value is the length to cut four wood dowels for the legs.

2c. Width of walker: Walkers are typically 635–735mm wide, but can be as narrow as 560–610mm if the user requires it to fit through narrow entryways. For a comfortable fit, make the walker slightly wider than shoulder width, or more if the user has a wider stride.

2d. Length of Top Front dowel: Subtract 35.5mm from the width value established above.

2e. Depth of walker: Proper depth allows the user's hands to fit comfortably within the handle. Ensure the handle is longer than the width of the user's fist, with extra room determined by the preference of the user.

2f. Length of Handlebar dowel: Add 82mm to the length of the desired handle.

2g. Length of Angled Front dowel: Perform the sine law by dividing the calculated length of the Top Front dowel by 2, subtracting 12mm, multiplying by sin(95.296°), and then dividing by sin(27.404°). Finally, subtract 75mm.

2h. Length of Side dowels: These are the last dowels cut, as dimensions depend on the slight variations during construction. Once the Ang Side connectors are placed in their proper locations, measure from one end of the circular stress reliever to the other, and then add 9mm to that value.

TABLE 2. Slicing parameters for PETG and TPU 85A filaments.

Slicing Parameter	PETG Value	TPU 85A Value
Layer Height	0.6mm	0.15
Wall Count	6	2
Infill Density	80%	30% (foot parts) 15% (handle)
Infill Pattern	Gyroid	Gyroid
Printing Temperature	225°C	238°C
Bed Temperature	85°C	50°C

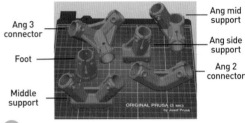

Ang 3 connector
Foot
Middle support
Ang mid support
Ang side support
Ang 2 connector

ORIGINAL PRUSA i3 MK3
by Josef Prusa

Ⓒ Orientation of the 3D-printed parts on the print bed.

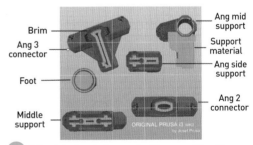

Brim
Ang 3 connector
Foot
Middle support
Ang mid support
Support material
Ang side support
Ang 2 connector

ORIGINAL PRUSA i3 MK3
by Josef Prusa

Ⓓ 3D-printed parts (orange) with support locations (light green) and brim (dark green).

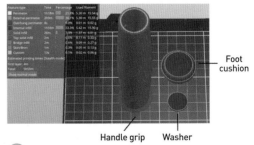

Foot cushion
Handle grip   Washer

Ⓔ Printing orientation of handle, foot cushion, and washer.

TABLE 3. Wood dowel parts.

Name	Number of Parts
Top Front	1
Handlebar	2
Leg	4
Angled Front	2
Top Side Support	2
Bottom Side Support	2

3. ASSEMBLE THE FEET

Insert the Foot Washer flush into the top of the Foot Body (Figure **F**). It should fit snugly.

Slide the Foot Cushion into the bottom of the Foot Body and exert a good amount of force to press it into the tight space.

Repeat for each foot.

> **TIP:** Press with your palm, which allows your full arm to exert force.

4. ASSEMBLE THE WALKER

4a. Measure and mark the center of the Top Front dowel. Then, mark half the length of the Middle Support part on either side of the first mark.

4b. Align the Middle Support through the dowel with the marks and secure it using wood screws and a drill/driver. You'll secure most, but not all, of the parts in this way as you go along.

4c. Measure and draw a center line along the length of the handlebar dowel (lateral area).

4d. Secure the Ang 2 Connector [L] and Ang 3 Connector [L] onto each end of the handlebar, with both stress relievers centered on the line and the letters right-side up (Figure **G**).

4e. Secure a Leg dowel onto Ang 2 Connector [L].

4f. Slide two Ang Side Supports [O] onto the Leg dowel with the stress relievers pointing to the right and the [O] symbol positioned at the top (Figure **H**).

4g. Secure a second Leg dowel onto Ang 3 Connector [L] on the end with the stress reliever pointing in the same direction.

4h. Slide one Ang Side Support [□] onto the Leg dowel with the stress relievers pointing to the right and the [□] symbol positioned at the top.

SAMPLE CALCULATIONS FOR PROPER FIT

Height of walker: Measured height of 860mm from the user's wrist to the ground.

Length of walker leg dowels to be cut:

$$\frac{Measured\ Height}{\cos\ 10°} - 15 - 30.75 = \frac{860}{\cos\ 10°} - 15 - 30.75 = 780\ mm$$

Width of walker: Desired overall width is 580mm.

Length of Top Front dowel to be cut:

$$Desired\ Overall\ Width - 35.5 = 580 - 35.5 = 545\ mm$$

Depth of walker: Desired length of handle is 150mm.

Length of Handlebar dowel to be cut:

$$Desired\ Handle\ length + 82 = 150 + 82 = 232\ mm$$

Length of Angled Front dowel:

$$\frac{(\sin\ 95.296°)\left(\frac{Length\ of\ Top\ Front\ Dowel}{2} - 12\right)}{\sin\ 27.404} - 75 =$$

$$\frac{(\sin\ 95.296°)\left(\frac{545}{2} - 12\right)}{\sin\ 27.404} - 75 = 488\ mm$$

Length of Top Side and Bottom Side Support dowels to be cut:

$$Measured\ Value\ From\ Ends\ Of\ Stress\ Relievers\ On\ Angled\ Side\ Supports + 9\ mm = 410 + 9 = 419\ mm$$

Stress reliever

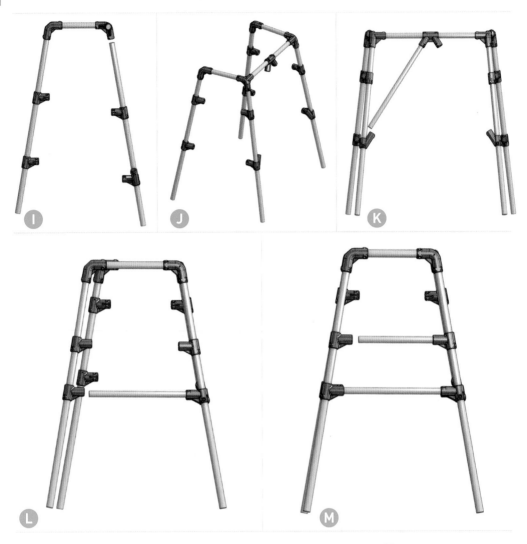

Following that by sliding the Ang Mid Support [L] (Figure ⓘ).

4i. Repeat steps 4c–4h for the right side, and where you used the [O] part you'll now use the [□] part. Insert (but do not secure) the last connection of the Ang 3 Connectors onto each end of the Top Front dowel. Now the overall frame of the walker is constructed (Figure ⓙ).

4j. Secure both Angled Front dowels into the Middle Support and lay the walker upside-down on a flat surface so the Top Front dowel side rests on the ground.

4k. Slide one of the Ang Mid Support onto the other end of the Angled Front Dowel. Be cautious when doing so, as the fit will be tight, and ensure the dowel sits fully into the

connectors (Figure ⓚ). Secure it once it's in place.

4l. Repeat step 4k for the remaining side.

4m. Secure the Ang 3 Connectors onto the Top Front dowel.

4n. Position the walker right-side up. Secure the Bottom Side Support dowel into the fixed Ang Mid Support part.

4o. Unsecure the Ang 2 Connector to add room and slowly move the corresponding Ang Side Support into the dowel (Figure ⓛ). Be cautious, as forcing it in place risks breaking the parts. Once fully in place, all the parts on that one side can be secured.

4p. Move the top Ang Side Connectors down from the top by 150mm. Insert the Top Side Support into one of the connectors (Figure ⓜ). Try to align both connectors in parallel and slowly move them toward the top of the walker so the Top Side dowel starts to fit into the other connector. Keep moving them incrementally until the dowel is fully in. Secure with screws.

4q. Repeat steps 4o–4p for the other side.

4r. Secure all four feet onto the ends of the Leg dowels (Figure Ⓝ).

4s. Go walking!

MOBILE UNLOCKED

This open source mobility walker has several user advantages:

- **Comfort and safety:** The inward angles of the front triangular bracing structure won't obstruct the legs of a sitting user; the A-frame creates rigidity by angling the front and back legs of the walker at 10° forward and backward, respectively, and angling the sides outward at 5°.

- **Weight:** The device mass was cut by about 20% (0.5kg or 1lb) compared to commercial walkers. Any difference in weight is significant for users with diminished strength, as static walkers require repetitive lifting when the user is in motion.

- **Strength:** All testing baselines demonstrate that the walker is robust — the user-applied load limit was found to be 375.3kg (827lbs) for vertical failure and 197.8kg (436lbs) for horizontal failure.

- **Affordability:** Finally, this open-source walker costs between $48–$68, which aligns with the least expensive commercial walkers or used walkers at a thrift store.

The cost could be brought down even further by substituting waste plastic for commercial 3D printing filament (see "PET Pultrusion," page 52), and by using alternate structural wood components — possibly reducing costs to less than 10% of the current.

More derivatives are also possible. When I had a broken foot, I used a single walker leg as a cane until it healed. Our lab at Western University is now testing a derivative of this design to make low-cost crutches! ⊘

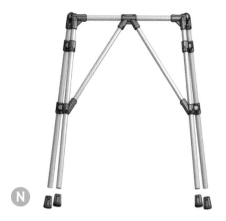

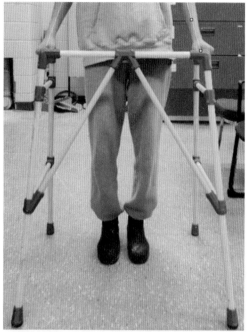

LEARN MORE:

- **UWO publication, with instructions and details on math:** doi.org/10.3390/inventions8030079
- **OSMS Project Library:** osms.li/movement-and-transport
- **Source code:** osf.io/v3njw (GNU GPL v3)

Details and links to everything: appropedia.org/Open-Source_Designs_for_Distributed_Manufacturing_of_Low-Cost_Customized_Walkers

Weather Bird

Make a science toy that forecasts the weather

Written and photographed by Bob Knetzger

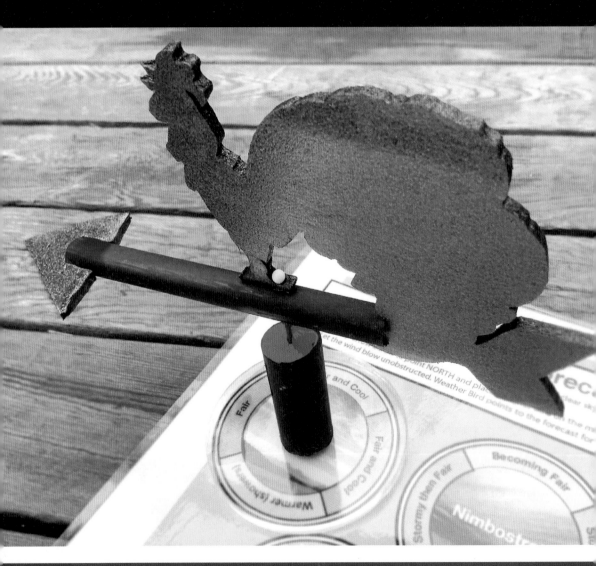

BOB KNETZGER is a designer/inventor/musician whose award-winning toys have been featured on *The Tonight Show*, *Nightline*, and *Good Morning America*. He is the author of *Make: Fun!*, available at makershed.com and fine bookstores.

The spinning Weather Bird wind vane points to tomorrow's forecast! But how does it know?

HOW IT WORKS

As warm and cold fronts cycle around low-pressure areas, they create likely sequences of clouds and wind directions. If you can figure out where you are in these swirling weather patterns, you'll have a pretty good guess as to what weather will come next.

In this toy, the panel of cloud wheels is designed from tables of observed clouds and wind directions. Look up: If you see high, wispy clouds with winds from the north, then fair and cool weather is ahead! Why? Because you're likely located west of a counter-clockwise-spinning low-pressure system with a cold front already past you. Keep in mind these tables are made for weather patterns in the Northern Hemisphere and can make only general forecasts. Predicting the weather is never an exact science — but you can try it yourself, learn about weather phenomena, and have fun by making your own Weather Bird!

(For more information and useful explanations about cloud types and weather fronts systems I recommend the Davis Instruments Weather Forecasting Quick Reference Card, Amazon B001444YUS.)

BUILD YOUR WEATHER BIRD

To make the Weather Bird wind vane, download the bird and arrowhead pattern from makezine. com/go/weather-bird. Print it out on heavy paper, trim and pin it to the foam tray, and then cut out the foam with a hobby knife (Figure ). (You can also use the downloaded SVG file to laser cut the paper bird pattern.)

Or, use the paper pattern as a guide with a hot wire foam cutter — it's really fun and you'll get a perfect result! See *Make:* Volume 16 or go online at makezine.com/projects/5-minute-foam-factory for complete plans to make a 5-Minute Foam Factory (Figures B and C). Lots of other foam cutting fun, too!

Of course, you don't have to make a fancy bird design, a simple arrowhead and fin shape will do.

Insert the foam bird cut-out in one end of the large straw (Figure D on the following page).

TIME REQUIRED: 1 Hour
DIFFICULTY: Easy
COST: Free

MATERIALS
» **Large plastic straw** like for bubble tea or a milkshake
» **Tiny plastic straw** such as a coffee stirrer, or a spray can nozzle straw like from WD-40
» **Polystyrene foam food tray**
» **Acrylic sheet, clear,** ¼" thick, 3" diameter
» **Straight pin**
» **Dowel or rod,** ¾" diameter, 2" length

TOOLS
» **Computer with printer**
» **Scissors**
» **Hobby knife**
» **Drill and tiny bit** to match your mini straw
» **Glue or wood screw**
» **Hot-wire foam cutter (optional)**
» **Laser cutter (optional)**
» **Laminator (optional)**

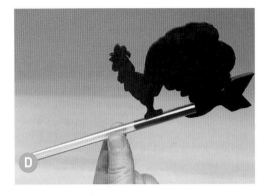

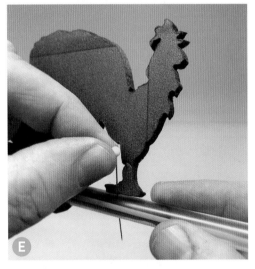

Carefully stick the straight pin through the top of the bird's foot, down through foam inside the straw, and out the bottom (Figure **E**). Be sure to stick the pin perpendicular to the straw. The pin is the vertical axle the bird spins on.

Insert the arrowhead on the other end so that it is horizontal and flat. Put the pin inside the small mini straw and check to see that the Weather Bird spins freely (Figure **F**). Blow on it to see that it points into the wind reliably. Trim the big straw if needed to balance.

Drill a small hole in the center of a 2"-long rod or dowel to accept the tiny straw (Figure **G**). Trim the mini straw to be slightly longer than the pin. Glue or screw the dowel to the center of a 3" circle of clear acrylic (Figure **H**). I painted the straw and dowel black to match.

Finally, download and print out the panel with the cloud wheels. Laminate it in clear plastic to be weatherproof if you like (Figure **I**).

USE IT

Go outside and look: What kind of clouds do you see? Find the closest matching cloud wheel on the panel and place the Weather Bird there. Line up the panel so it points north. When the wind blows, Weather Bird points to the forecast!

Keep a daily log of your Weather Bird's predictions and compare them to the official weather forecasts and the actual weather. How did the Weather Bird do? ✪

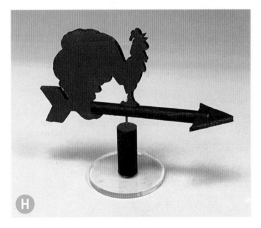

FUN FACT!

Why roosters on weather vanes? It's from a papal decree in the 9th century that said every church should have a rooster on top because Christ told Peter, "This very night, before the rooster crows, you will disown me three times." To this day, the decorative birds are found on barns, steeples, and rooftops.

1. Look at the sky and find the closest matching cloud wheel. (If clear sky or less than 30% clouds, then no change in the weather.)
2. Line up arrow below to point NORTH and place Weather Bird on the matching cloud wheel.
3. Let the wind blow unobstructed. Weather Bird points to the forecast for the next 12-36 hours.

Point NORTH

Cirrus — Fair and Cool, Fair and Cool, Warmer (showers), Fair

Nimbostratus — Becoming Fair, Stormy then Fair, Becoming Fair, Stormy then Fair

Altostratus — Fair and Cool, Cloudy, Becoming Fair, Showers

Cumulus — Showery, Cool, Showery, Showers then Fair, Fair

Stratocumulus — Unsettled, Little Change, Showers, Little Change

Altocumulus — Little Change, Little Change, Showers, Fair

3D PRINTING
CERAMICS

Going from the 3D printer to the kiln

Written by Joan Horvath and Rich Cameron

A WASP clay printers: model 40100 (left) and 2040 (center).

Ceramics are an ancient material, but we think of them as high-tech when it comes to 3D printing. However, as with many 3D printing techniques, what used to be a high-tech industrial process is starting to become accessible to the garage enthusiast. And if you're already creating and firing ceramic pieces, you may have a lot of what you need already.

There are two fundamental types of 3D printable ceramics. In one case, **clay is directly printed** with a special-purpose paste printer.

In the other, you print "ceramic" filament with **clay particles embedded in a plastic carrier or binder,** and you need to "debind" out the plastic before firing the piece. There is also a (challenging) resin printer version of this latter method.

In all these scenarios, a standard kiln will work for the firing stage. This is in distinction to the metal-infused filaments we described in our piece in *Make:* Volume 86, "Lower Cost Metal 3D Printing." Those require a higher-tech furnace with tight control of temperature for sintering, and often vacuum or an inert gas during some of the process. For these ceramic applications, the kiln just has to reach high enough temperatures with a controlled profile. For all the options described in this article you can learn more about firing on the company's websites.

Design issues with printed ceramics are similar to those with metal filament, like allowing for shrinkage during debinding and sintering. For some materials, gravity will make the vertical, z-axis direction shrink a bit more than the other two, but this is not universal.

CLAY 3D PRINTERS

Let's talk first about directly printing in clay, which stays pretty soft out of the printer. This type of printable ceramic can be worked with like any other clay, adapted a bit by hand if desired, and then fired in a kiln. Italian 3D printer manufacturer **WASP** (3dwasp.com) sells its Delta WASP 2040 Clay printer (Figure **A**) for €3,300 ($3,588), and the larger 40100 model for €7,590 ($8,252). This method of printing clay also requires a compressed air supply; you'll need an 8 bar (116psi) compressor with a 50 liter (13gal) tank. European compressor options are specified at WASP's technical specs pages.

These delta printers are aimed at the market for creating larger ceramic pieces (Figures **B** and **C**) — the 2040 model can print parts up to 20cm in diameter and 40cm high, and the 40100 can handle 40cm diameter by 1 meter high. The main negative to this approach is buying a special-purpose printer, though it is also possible to buy clay retrofit kits for WASP's more general-purpose deltabot 3D printers.

Obviously you aren't going to buy a specialized clay 3D printer unless you are thinking about doing a lot of this sort of thing. However, if that is your intent, the material for this printer is vastly cheaper than clay-infused filament. Red earthenware clay is €15 for 12.5kg, or a little over €1 per kg. Other materials will vary in price, and shipping from Italy will need to be factored in for customers on other continents.

WASP has online courses available in English and Italian to get you started with clay. The products created are fired in the usual way

B WASP clay print with complex geometry.

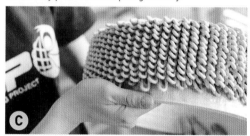
C WASP clay print with a yarn-like texture.

WASP

for this type of clay, without the requirement for specialized kilns. One plus is that since it is soft coming off the printer, some traditional clayworking and smoothing can be done before firing. Prints will also shrink a bit during firing; how much will depend on a variety of conditions.

There are several other entrants in this space, like 3D Potter in Florida, VormVrij in the Netherlands, and Stoneflower in Germany. It's a dynamic market, so if you're thinking about a purchase, do a bit of reading reviews and considering shipping costs and material availability in your tradeoffs.

CERAMIC-INFUSED FILAMENT

An alternative method is to use a ceramic-infused filament, and then bake out the plastic carrier after printing. The tradeoff here is that you don't have to buy a specialized 3D printer, but the material is considerably pricier than the bulk clay used by paste-style printers. These ceramic filaments are also going to be abrasive, so swapping a hardened nozzle into your printer is a must.

There are various materials emerging in this market. Some are aimed at specialized engineering markets that want to create delicate

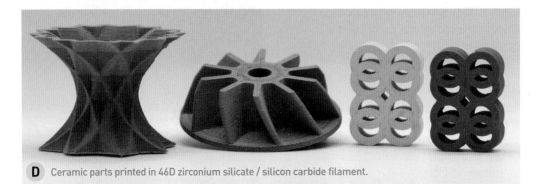

D Ceramic parts printed in 46D zirconium silicate / silicon carbide filament.

parts with particular properties that can't really be formed with conventional techniques. Others are designed to work with a low-end consumer 3D printer, and then baked out in a (temperature-controllable) ceramic kiln. Needless to say, parts will shrink substantially as the binder bakes out or is removed chemically, although the manufacturers say that the shrinkage is fairly symmetrical on all axes.

The **Virtual Foundry** (thevirtualfoundry.com) has a product called Ceramic Clay **Filamet** that works this way. It does not require any separate debinding chemicals, just a heat-debinding process and then firing in a kiln. Here are costs (including a basic 3D printer) from Virtual Foundry for what it would take to get started printing clay (**Table 1**). They also have a Filamet version of Pyrex (borosilicate) glass.

Part feature size (print resolution) is consistent with what you would expect from a consumer 3D printer. The parts shown in Figure **D** are made of 46D zirconium silicate and silicon carbide. In Figure **E** we see a sintered 46D part perched on an unsintered part of the same material; the sintered (lighter) piece is about 2" across at the base. The company notes that it is relatively easy for them to make customized mixes if a customer needs something unique.

French ceramics manufacturer **Nanoe** (nanoe. fr) has several types of ceramic-infused filament — white and black zirconia, silicon carbide, and alumina — which it sells under its **Zetamix** brand (zetamix.fr). These are mainly intended for the specialty technical ceramic market, costing in the €200–€400 range for half a kilogram, with interesting dielectric or other properties.

The Zetamix debinding process requires

TABLE 1: FIRST PART COST

3D printer — Ender 3 S1	$322.00
Hardened steel nozzle, 0.6mm	$13.52
Filawarmer module	$85.00
Build plate prep: blue painter's tape	$7.88
Amaco 46-D Ceramic Clay Filamet, 1.75mm (0.25kg)	$63.50
Alumina crucible, 300ml	$60.00
Sintering refractory ballast: magnesium 1.75mm (0.25kg)	$35.04
Kiln — FireBox 8x4 LT Ceramic Kiln FK	$1,072.70
TOTAL	**$1,659.64**

SECOND PART COST

Amaco 46-D Ceramic Clay Filamet, 1.75mm (0.25kg)	$63.50
TOTAL	**$63.50**

E Sintered 46D resting on unsintered.

The Virtual Foundry, Nanoe

soaking the part in acetone as a first step, which removes some of the binder. Then, after the acetone has been allowed to evaporate from the part, the remaining binder is baked out in the kiln. Since acetone is flammable, this process requires careful attention to the manufacturer's process and knowledge about how to safely ventilate and manage acetone around a kiln. The Zetamix website has extensive documentation for each material on design limitations and process guidelines.

PORCELAIN FILAMENT

Nanoe's newest entry in the line is a filament that can be used to create porcelain parts, **Zetamix Porcelain**. Porcelain is denser than other ceramics and less porous. Traditionally made with a high proportion of kaolin clay, it can be made thin and translucent. Porcelain has been used for centuries to make elegant, thin, but watertight pieces like fine teacups. It also holds paint and glazes well.

While still pricey at €99 for a 500g spool (Figure **F**), the porcelain filament does not require buying a specialized printer or kiln beyond what one would otherwise use for these ceramics. Parts will, however, shrink more in the z (vertical) axis than the horizontal ones; it's recommended to print them at 116% scale in x and y, and 136% in the z axis.

After bake-out steps like those for the other Zetamix ceramics, the porcelain can be glazed and fired (Figures **G** and **H**) which will largely fill in the layer lines, leaving a fine surface (with, of course, some texture remaining). This is a new area for experimentation, ripe for finding innovative ways to mix new and old techniques.

CERAMIC RESIN

Rounding out the options for consumer-level printing of ceramics are **resin ceramics**, although admittedly this option may be challenging for most. **Formlabs** (formlabs.com) has a technical ceramic resin, Alumina 4N (Figure **I** on the following page), available only for use on its Form 3+ printer; it requires a compatible build platform and tank as well. The intended use cases are high-temperature, high-strength, intricate engineering parts that would be difficult

F Zetamix Porcelain filament spool.

G Glazed porcelain parts 3D printed with Zetamix Porcelain filament.

H Glazed porcelain parts being fired.

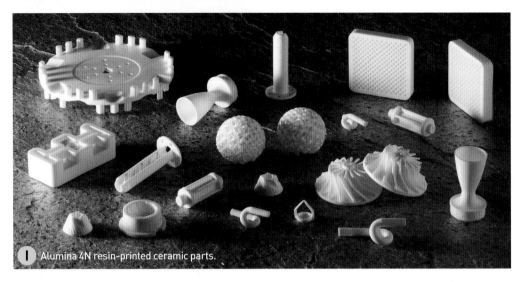

Alumina 4N resin-printed ceramic parts.

to fabricate in other ways, and that can withstand corrosive environments. As always, resin printers shine for tiny parts requiring fine features.

At $1,299 for a liter, this resin is not intended for hobbyist use. (In fact, the Formlabs store page for Alumina 4N makes you check a box acknowledging that you will need more equipment and experience before it lets you put it in your cart.) The resin has ceramic particles mixed into it, which is obviously challenging for a technology that relies on illuminating resin to cure it. The abrasiveness is problematic as well, since SLA printers need to keep their trays clear and free of scratches to perform well. Bottom line: not for SLA printer beginners.

In addition, parts printed with this material have to be washed in a unique Ceramic Wash Solution since they will crack in the usual

Crane WASP 3D printer, print volume 6.3m diameter × 3m high (approximately 21×10 feet).

Formlabs, WASP

isopropyl alcohol used for cleaning, or for that matter, in water. Then parts must be thoroughly dried and then fired. Formlabs predicts shrinkage of 22% in the horizontal plane and 26% in the vertical. If you're thinking about it and own a printer, read Formlabs' documentation of the process and printer settings before embarking. And of course you'll need to obtain a kiln capable of maintaining the required temperature profiles.

If you are very experienced and have a compatible Form 3+, and you need a technical part that can't be formulated with one of the other ceramic techniques we've discussed to this point, this might be your only 3D printing option.

GOING REALLY BIG

Now that we've progressed from direct-printing pieces on the order of half a meter tall down through delicate technical pieces, let's consider what happens if you want to create something on the scale of a house. Creating adobe and brick dwellings is arguably the oldest additive manufacturing, and you might ask whether ceramic printing techniques can scale up to create house-sized objects.

Printer company WASP, mentioned earlier in this article, has created an experimental 3D printer (Figure **J**) to **create buildings** out of locally available materials, like one they have dubbed their Gaia house (Figure **K**). Their goal was to use readily available **soil, lime, and waste fiber** from agriculture, converted into building materials. You can read more about it at 3dwasp.com/en/3d-printed-house-gaia. They have since gone on to accept building challenges in other materials, and it will be interesting to see whether concrete or clay-like 3D printing eventually wins out in this space.

THE BOTTOM LINE

3D printing ceramics with consumer-level equipment is not only possible, but a vibrant space with a growing number of competitors. We have given a representative sampling here, but if you have an application you want to try, you should do a bit more searching to see which of these options makes sense for you.

From an economic point of view, if you imagine yourself doing a lot of parts and extremely fine

K Gaia house, printed by Crane WASP printer from local soil, lime, and ag waste fiber.

detail is less critical, a special-purpose paste printer is probably the way to go. The material will be cheaper than filament by one or two orders of magnitude, and it may be easier to do your own material experimentation. You'll need to invest in a printer, kiln, and peripheral equipment like a compressed-air source.

Filament lets you get started more cheaply, particularly if you already have a suitable printer (with a hardened nozzle) and kiln. What type of filament you use will be driven by the precise material you want, and what type of debinding and post-processing you feel comfortable doing.

Resin ceramic printing is likely to remain the domain of industrial applications for a while, although one never knows in 3D printing when something innovative will come down the pike.

And of course we hope that our friends at WASP are indeed at the forefront of a trend to build substantial, attractive housing with local materials where people might otherwise be in tin-roofed shacks at best. We hope to hear more innovative applications of ceramic 3D printing at all scales in the near future. ❷

JOAN HORVATH & **RICH CAMERON** are the co-founders of Nonscriptum LLC (nonscriptum.com). They are the authors of many books, including *Make: Geometry*, *Make: Calculus*, and *Make: Trigonometry*.

FINE FINISHING
FOR FUSED FILAMENT

Getting that silky smooth surface finish on 3D prints

Written and photographed by Elyse Farris

ELYSE FARRIS is a cosplayer and propmaker who loves to share DIYs on Instagram and TikTok. @leesymaecosplay

3D printing your own models and props is a fun hobby that takes you on the adventure of learning ... that there's a lot of work involved after the print reaches 100%.

We're going to highlight a few ways of smoothing out 3D prints once they're finished printing. It's easy and you only need about $50 to get set up to finish lots of prints.

ELIMINATING LAYER LINES

Why finish or "smooth out" your prints? Most printers, even at their finest settings, will still leave layer lines. You'll be able to see these layer lines if you just paint over the texture.

A great way to start is with a mixture of Bondo Spot Putty and ordinary acetone. Bondo Spot Putty is creamy, and great for filling small defects. It's also very easy to sand. To make it even thinner for filling something as small as layer lines, just add a little acetone to thin out the putty to form it into a paint-like substance. Paint it on with a disposable brush. It will dry in about half an hour or less (Figures **A** and **B**).

If you have any small defects left that didn't get filled, use the putty again and sand smooth (usually up to 400–500 grit) once it's dried.

SMOOTHING SEAMS

If your prop is too large for the printer to print in one piece, you may end up welding your prints together for a strong bond. Welding is great, but can create ugly seams, like on my X-wing helmet on the opposite page. Or maybe there was a layer shift during printing and it caused a weird seam to appear.

This can be fixed with Bondo Body Filler (Figure **C**). It's a gray, two-part resin-based filler that's very thick and dries super strong. (You can see it applied over the orange Spot Putty coat on the helmet.) You'll probably want an electric sander to smooth this down. It isn't necessarily good for layer lines because it's so thick.

HIGH-GRIT SANDING

So now that any seams are hidden and layer lines are smooth, give your print a coat of Filler Primer (Figure **D**). It's basically sandable spray paint.

Once it's dry, this is where high-grit sanding comes in. Start with 400 and go all the way up

TIME REQUIRED 1–7 Days

DIFFICULTY Easy

COST $50–$60

MATERIALS
» Acetone
» Bondo Glazing & Spot Putty
» Bondo Body Filler
» Automotive 2-in-1 filler/primer, sandable

TOOLS
» Sandpaper, 200–1000 grit
» Disposable brushes
» PPE: Goggles, respirator, gloves
» Electric sander (optional)

A

B

C

D

to 1000 or more if you desire. Once the prop is completely smooth it's ready for primer and paint!

NOTE: You should always wear a respirator and work in well-ventilated areas while sanding.

This is just the way I do it. You'll find that there are dozens of methods for smoothing 3D prints. All of these methods work as long as you're happy with the end result! ⊘

TIM DEAGAN
makes, breaks, and
collects things in
Austin, Texas. He loves
the experience of using
software to breathe
life into hardware
with Linux and any
microcontroller he can
get his hands on.

Written and photographed by Tim Deagan

SPEED WEED!

How to design vinyl cutter projects for fast, easy weeding

3D printers, CNCs, laser cutters, and other devices that move a toolhead around under digital control have become a major focus of maker activity over the last couple of decades. Often lost in the discussion of digital fabrication is a tool with a massive and extremely active user base: *stencil cutters* (Figure A). Often referred to as *vinyl cutters* due to the most commonly cut material, these tools are sold by the millions to makers, crafters, scrapbookers, DIYers, and professionals. Desktop craft cutters like Cricut and Silhouette have been sold in retail outlets far

longer than 3D printers. Wide-format vinyl cutters have been a mainstay of sign-making companies around the world.

The tool that these machines move is technically referred to as a *drag knife* — a sharp pointed blade that can swivel freely around its vertical axis. When the stepper motors (or servos) change direction, they drag the blade. The tip, which is slightly off-center from the axis of its shaft, hangs back enough to reorient the blade along the new line of motion. This swiveling motion, combined with the machine's ability to raise and lower the blade, allows you to precisely cut out amazingly complex patterns. Adhesive or heat-bonded vinyl is extremely popular, but paper, thin wood, leather, and other materials thin enough for the machine to handle are also commonly cut.

Unlike a traditional stencil, where **bridges** are made to make sure the whole image is of one piece, stencil cutters can take advantage of the adhesion of the backing material to allow the design to have many independent pieces. Some of these pieces then must be removed so that the desired pattern can be transferred to the target surface (mug, vehicle, wall, whatever). The process of removing these unwanted parts of the cut image is referred to as **weeding** (Figure **B**).

ENTER SPEED WEEDING

Weeding can be a tedious, frustrating experience if the pattern is complex and has many tiny, oddly shaped pieces or long intertwining sections to remove. Vinyl sticks to itself with a passion, so if you accidentally let a piece you're removing touch a piece you want to stay, it can be next to impossible to separate them. It's not uncommon to get an hour into a weeding session and have a piece drape out of your fist, stick to the design, and pull up pieces that need to stay, requiring the whole thing to get thrown out and started over.

There are lots of great videos and tutorials online to help folks improve their weeding experience. One of the most effective techniques is **speed weeding**. Speed weeding typically consists of adding extra cuts to your design in advance, to reduce the length of long trailing sections that will need to be weeded. Combined with a few tricks from traditional stencil making to reduce independent **islands**, some thoughtful effort in the design stage can make a huge difference in the experience of weeding your patterns.

FIRST THINGS FIRST

Whether you're using a small desktop cutter or a wide-format commercial cutter, you must set up your machine correctly for the material you're cutting. For example, if you're cutting a material like vinyl that has a paper backing material, or you've adhered your material to a reusable

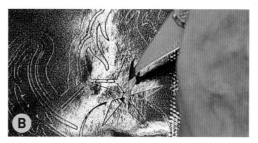

A 9" Silhouette SD craft cutter atop a 34" US Cutter MH 871-MK2 vinyl cutter.

Weeding unwanted pieces from a design.

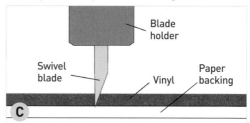

Blade depth: Set it to cut the vinyl but not the backing.

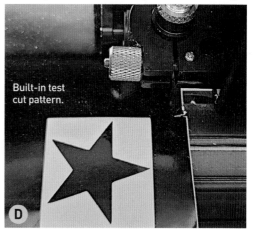

Built-in test cut pattern.

backing sheet, your blade depth must be set up to fully penetrate the target material and barely scratch the surface of your backing material (Figure **C**).

Your machine will provide instructions on how to do this calibration. Most machines provide a quick "test cut" capability that will let you verify that everything is set properly (Figure **D**).

E Original image for window sticker.

F Vinyl with pattern cut into it prior to weeding.

ORIGINAL

WITH BRIDGES

G Letters before and after adding bridges.

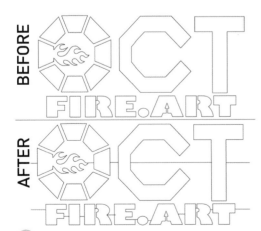

H Vector image with and without speed weeding cuts.

Besides the blade depth, the travel speed and blade pressure are typically adjustable as well. Some materials require different types of blades; this is usually related to the angle of the blade needed for the desired depth of cut. Follow your machine instructions and use the test cut function liberally. It's far better to waste a little material on a small test cut than to produce a long, complicated pattern and have to throw it out because the cuts weren't effective.

THINKING ABOUT NEGATIVE SPACE

I've recently been making large runs of window stickers to give out at festivals (Figure **E**). This requires a lot of weeding. To speed the process up, I visualized how I wanted to remove the waste in the fastest way possible. This meant trying to pull strips of vinyl that wouldn't catch, or need to reverse direction, or be so long that they'd drape and catch on the pieces that had to stay.

Staring at a piece of work hot off the cutter can be daunting. It looks a lot like an uncut piece of vinyl with a bunch of scratches on it. The beautiful image you've cut is hidden among all the waste material that must be removed (Figure **F**). The act of weeding is the creation of the negative space in your image, like removing stone from a slab until the sculpture emerges.

Thinking about how this negative space will be created, the way in which you'll remove the waste pieces, is the ticket to making your weeding life easier.

The three key ideas to making your design better for speed weeding are:
1. You can modify your image to be more stencil-like by making bridges,
2. You can add as many cuts into the negative space as you like, and
3. You can make cuts into or through the remaining material of your image without it being a problem when you adhere it.

Let's look at each of those ideas and how they might be put to use.

ADDING BRIDGES

My original image had a couple of tedious bits to remove if I was going to make a couple hundred of these stickers. The holes inside the R and the A, called **eyes** by typographers, would need to be carefully picked out. I didn't want my entire font to be a stencil font, so I used InkScape to create bridges from the outside to the eyes only (Figure **G**). Working in a **vector drawing tool** like InkScape, Adobe Illustrator, or possibly the software provided with your cutter is essential for effective modification and preparation of images for cutting.

ADDING EXTRA CUTS

Having eliminated the fiddly little bits in the letters, I wanted a no-hassle set of weeds to pull. After some experimentation, and there's no substitute for making test cuts to try things out, I determined that four pulls would be the easiest to reliably make (Figure **H**). Also, I really wanted to cut a grid of these, so the cut lines are intended to match up to duplicate images above, below, and to each side so that I can weed multiple stickers all at once (Figure **I**).

Placing these additional cuts is usually as simple as adding straight lines in your vector editing tool. These lines do not need to connect their nodes to other lines in the drawing. The cutting tool software will simply execute these straight-line cuts on top of whatever else is there.

I will usually add these additional cut lines on a separate **layer** so that I can turn them on and off as a set. If you're not using layers in your graphics tools, you'll want to add them to your skill set — they make management of your images dramatically cleaner (Figure **J**).

CUTTING INTO THE BASE IMAGE

When you're adding speed weeding lines to your image, you almost always want them to cross some other cut. If there is a gap between lines, the tiny bit of vinyl that isn't cut can be surprisingly strong, and instead of breaking where you want, it will pull additional vinyl along with it (Figures **K** and **L** on the following page). To avoid these tiny tabs, you want to make sure that the pieces you're weeding are completely cut all the way around.

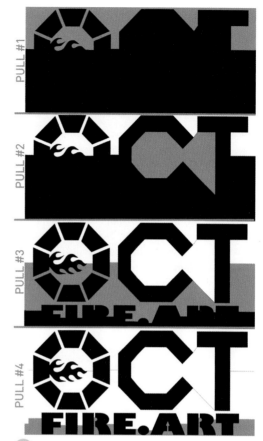

PULL #1

PULL #2

PULL #3

PULL #4

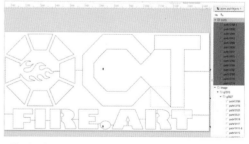

I Sequence of weeding pulls, each pull shown in blue.

J InkScape graphics file with highlighted speed-weed layer.

K Pulling too much vinyl when weeding.

L Close-up of speed weeding lines not reaching to image lines, causing undesirable tabs.

The way to do this is to extend your cutting lines slightly into the material you aren't pulling away. This may sound untidy at best and sloppy at worst. The thing to remember is that once the final sticker is applied, small extra cuts in the material are almost always invisible. Even sizable cuts that divide a remaining piece in two aren't necessarily a big deal.

PULLING IT ALL TOGETHER

While laser cutting, CNC milling, and other digital fabrication processes all rely on essentially the same motor-driven tool approach, each has its own unique aspects and techniques that impact workpiece design and practical implementation. Vinyl cutting, whether you're using a 34"-wide cutter or a little desktop Cricut or Silhouette, can be tremendously tedious and frustrating unless you use ideas like speed weeding to make your life easier.

As you work with your stencil cutter designs, practice picturing what weeding out those unwanted pieces will entail. I feel confident you'll rapidly adopt speed weeding as a method of anger management in your efforts. And once you do, everything you look at will start needing stickers! ◐

Friends helping apply a heavily weeded vinyl sticker to my RV.

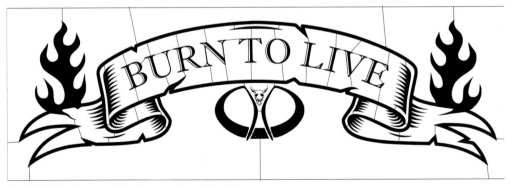

The RV sticker's image with added speed weeding cuts.